W9-ATD-600

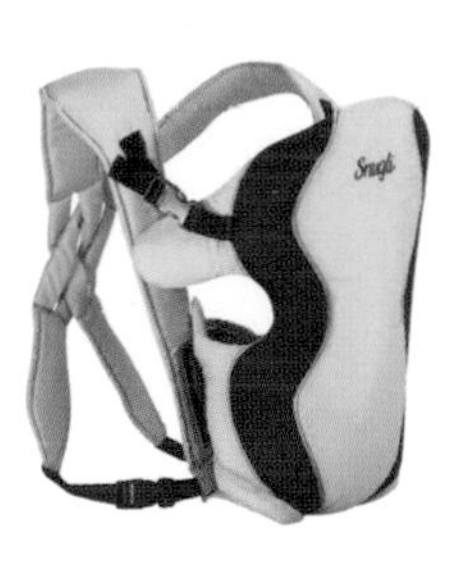

WOMEN OF INVENTION

KOTEX FEATHERWEIGHT
SANITARY BELT
NON ELASTIC
WAR DURATION TYPE
HEATER

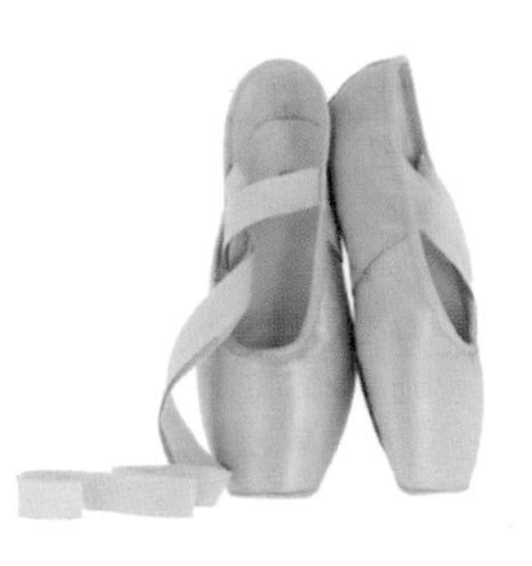

WOMEN OF INVENTION

LIFE-CHANGING IDEAS BY REMARKABLE WOMEN

CHARLOTTE MONTAGUE

CHARTWELL
BOOKS

CONTENTS

Key: **Technology** **Science** **Engineering** **Medicine** **Transport & Communications**

Beauty & Fashion | Culinary & Food Technology | Domestic Technology | Healthcare | Toys & Games

Key: Technology | Science | Engineering | Medicine | Transport & Communications

Beauty & Fashion | Culinary & Food Technology | Domestic Technology | Healthcare | Toys & Games

INTRODUCTION

Machines, drugs, hair treatments, ice cream makers, aircraft, central heating systems, home security systems ... women have invented things in a staggering variety of fields, often working in obscurity or with little chance of having their names attached to their discoveries, let alone gaining recognition or actually making any money from them. Who knows how many women inventors remain unacknowledged or unknown? For so long in the eighteenth and nineteenth centuries, the system would not even allow female ingenuity to be rewarded by the granting of a patent in a woman's name. Instead, the certificate was issued in the name of her husband or some other male.

If the woman happened to be African American, she had no chance at all. Black female inventors had to resort to selling the rights to their creations to white men for negligible amounts of money. Often those men went on to profit greatly from the ingenuity of another person's mind. There was also the prospect, if a product was invented and marketed by a black woman, that racially prejudiced white women would be reluctant to purchase it. Madam C.J. Walker was an exception, and led the way for female African American entrepreneurs across America.

Those have not been the only difficulties for women inventors, however. Problems created by the gender politics of the eighteenth, nineteenth, and twentieth centuries can be added. If a woman was interested in a scientific career in the nineteenth or twentieth centuries, for instance, it was impossible on some occasions to be admitted by a university and even once there she would often be prohibited from attending lectures. It is shocking to learn of the reluctance on the part of some eminent scientists to allow women into their lecture theaters or laboratories.

In the workplace too, women inventors experienced startling difficulties until very recently. When Patsy Sherman, inventor of Scotchgard, was working at the huge multinational 3M in the early 1950s, women were banned from entering the textile mill where products were tested. So, even though Patsy had created the product that was being tested, she was not allowed to be present. She was left outside pacing nervously up and down awaiting the results.

In some ways World War II provided opportunities for many women in the scientific and technological fields. Men were going off to fight overseas, leaving gaps in the staffing of the workplace. Many women jumped at the chance to work in the scientific arena and many retained their jobs after the war, creating new drugs, new machines. At last, they began to change the place of women in society. Hedy Lamarr proved she was

Madam C.J. Walker's beauty and hair products made her one of the most successful African American business owners ever.

more than just a pretty Hollywood film star when she patented a highly complex frequency-hopping communication system in 1942.

It remained tough, however, and there are several instances of women missing out on recognition for their contributions to advances in human knowledge. One of the most disappointing examples is that of the English scientist Rosalind Franklin, instrumental in the discovery of the structure of the DNA molecule but who was simply airbrushed out of the story as two men, Francis Crick and James Watson, took all the glory as well as the 1962 Nobel Prize in Physiology or Medicine.

Another woman who lost out was the Chinese-born American experimental physicist Chien-Shiung Wu who was not even named as a joint winner of the 1957 Nobel Prize in Physics although she had played a fundamental role in the work that won it for her colleagues, Professor Tsung-Dao Lee and Professor Chen-Ning Yang. Wu had worked on the Manhattan Project and in later years became known as "the First Lady of Physics."

Despite all of these problems, hindrances, prejudices, and disappointments, women have proved every bit as ingenious as their male counterparts and perhaps even more so, as they have had to work harder to gain any recognition at all. Some like Marie Curie have become world famous, others such as Frances Gabe, inventor of the self-cleaning house, have died in obscurity.

The women in this book come from all walks of life. Needless to say, there are examples of domestic, culinary, homeware, and fashion inventions by women, as those are the areas to which most women were restricted. By contrast there are aeronautical engineers such as Beatrice "Tilly" Shilling, the English woman whose solution during World War II to a problem with fighter aircraft engines helped the RAF win the Battle of Britain. Spectacularly talented women such as the French flier, sportswoman, and nurse, Marie Marvingt, have amazed us with their ingenuity, courage, and sheer determination. But then there are also entrepreneurs such as the Canadian-American Martha Matilda Harper, who came from nothing to establish an international empire of franchised hair salons. There are also women whose work has changed the world and saved countless lives, women like Virginia Apgar, creator of the Apgar score by which the health of newborn babies is measured and monitored.

As this book proves without a doubt, invention is far from a male preserve, and we should be endlessly grateful for the efforts of all the women assembled within these pages to be original and innovative, often with the odds totally stacked against them. You don't need to be a specific gender to be a great inventor, all you need according to Beulah Louise Henry is "time, space and freedom." Or as Patsy Sherman once said, "Girls should follow their dreams. They can do anything anybody else can do."

Maya Burhanpurkar is a science prodigy with a great future ahead of her.

RANDI ALTSCHUL

THE WORLD'S FIRST DISPOSABLE CELL PHONE INVENTION: THE PHONE-CARD-PHONE

Often we think of throwing our cell phones against the wall in frustration when we lose signal in mid-conversation. With the phone invented in 1999 by Randi Altschul, you could actually do just that. It was the world's first disposable cell phone.

Randice-Lisa "Randi" Altschul was no stranger to new ideas. Born in 1960, she was a toy designer who had made a fortune by the age of 26 from inventing toys and games. Her invention of the Phone-Card-Phone shows that you do not need to be an expert in a field of technology to create something new and innovative in that area. Altschul had a unique approach, she said:

> *The greatest asset I have over everyone else in that business, is my toy mentality. An engineer's mentality is to make something last, to make it durable. A toy's lifespan is about an hour; then the kid throws it away. You get it, you play with it, and—boom—it's gone.*

The phone was made from recycled paper products and its body was the size of a credit card, only three times thicker. It used an elongated flexible circuit that was part of the phone's body, the ultra-thin circuitry was created by applying metallic conductive inks to the paper.

Altschul saw her disposable phone as invaluable to parents who wanted their children to be able to call them at any time to ensure their safety. Sadly, other telephone companies knew a good idea when they saw one and began to market their own disposable small phones. Undeterred, Randi Altschul continues to invent and is working on other ideas, one of which is a disposable laptop computer. She is unashamed in her ambition, saying:

> *I'm going cheap and dumb. In monetary terms, I want to be the next Bill Gates.*

The Phone-Card-Phone is a real cell phone which makes outgoing calls only. It has 60 minutes of calling time and a hands-free attachment.

BETSY ANCKER-JOHNSON

CHANGING ATTITUDES TO WOMEN IN SCIENCE
INVENTION: THE GIGACYCLE RANGE SIGNAL GENERATOR

Born in 1927 in St. Louis, Missouri, Betsy Ancker was encouraged by her mother to pursue whatever was of most interest to her. She chose physics, enrolling at Wellesley College, the famous women's college in Boston, from which she graduated in 1949 with a bachelor's degree. She moved on to the University of Tübingen, in Baden-Württemberg, Germany, where she gained a PhD in physics in 1953. In 1958, she married Hal Johnson and they had four children.

In 1956 she became a Senior Research Physicist for Sylvania Electric Products at their Microwave Physics Laboratory in Palo Alto, California, and began to specialize in plasma which she described as a fourth state of matter after liquids, solids, and gases. She wrote a number of papers on the subject of plasma and her work led to the finding that solid-state plasmas can be used as microwave sources of radiation.

In the 1960s she worked at Boeing Research Laboratories in Seattle, where she discovered that if a low-density plasma is established in a piece of semi-conductor material in the presence of a high-intensity electric field and low-intensity parallel magnetic field, high frequency signals well into the gigacycle range can be generated. Her Gigacycle Range Signal Generator transmits repeating and non-repeating electronic signals used commonly in microwaves, arbitrary waveforms, and pitch.

In 1973, Ancker-Johnson became Assistant Secretary for Science and Technology at the US Department of Commerce and two years later, she became just the fourth woman elected to the National Academy of Engineering. She was the first woman ever to occupy the position of vice-president in the automotive industry when she took on the role of Vice President of General Motors' Environmental Activity Staff in 1979.

Betsy Ancker-Johnson was 90 years old in 2017, she has published over seventy scientific papers and patents. She is one of a generation of women who helped inspire a reassessment of attitudes to women in science and industry.

WHAT IS PLASMA?

The term "plasma," meaning "moldable substance," was coined in the 1920s by the Nobel Prize-winning American chemist Irving Langmuir. It is one of the four fundamental states of matter, the others being solid, liquid, and gas, although it does not exist naturally on earth and can only be created artificially from neutral gases.

A plasma is an ionized gas into which enough energy is provided to free electrons from atoms or molecules, and to allow both ions and electrons to coexist. To the best of our knowledge, plasmas are the most common state of matter in our universe.

Gases can be transformed into plasmas in a variety of ways, but all of them involve pumping the gas with energy. For instance, a spark in a gas will create a plasma and a hot gas passing through a big spark will make that gas a plasma that can be of use. Thus, plasma torches can be used in industry to cut metals. The sun, by the way, is nothing more than a huge ball of plasma.

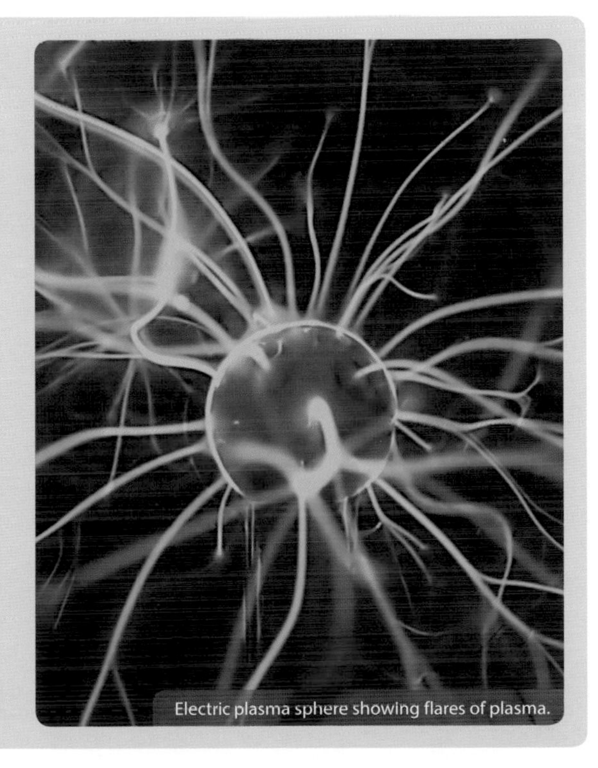

Electric plasma sphere showing flares of plasma.

THE NATIONAL ACADEMY OF ENGINEERING

One of the greatest honors an American engineer can receive is election to the National Academy of Engineering. Founded in 1964, it operates engineering programs that are aimed at meeting national needs, encourages education and research, and recognizes the achievements of America's engineers.

Its first female member was Lillian Gilbreth, the psychologist and industrial engineer who was elected in 1965. The second woman elected was the brilliant computer scientist Grace Hopper, who received the honor in 1973, and the third inductee was the so-called "queen of carbon science," Mildred Dresselhaus, elected in 1974. Betsy Ancker-Johnson was the fourth. As of 2017, the NAE had 175 female members (US and foreign, not including deceased) out of over 2,000 active US members and over 200 foreign members.

Lillian Gilbreth (1878 – 1972), a distinguished industrial psychologist, engineer, and mother of twelve, was the original super-mom. Although an accomplished professional credited with many academic and industrial "firsts," including kitchen and household appliance designs, it was as the role model for the highly intelligent, quirky mother in the popular movie *Cheaper by the Dozen* that she is probably most widely known.

MARY ANDERSON

A CLEAR VISION OF THE WAY AHEAD
INVENTION: THE WINDSHIELD WIPER

Back at the start of the twentieth century, rain on the windshield of a moving vehicle was just something you had to live with. The only way to deal with it was to stop the car every now and then and wipe the glass with a cloth. It took a woman named Mary Anderson to do something about that problem.

Born in Greene County, Alabama, in 1866, the year after the end of the American Civil War, she moved in 1889 with her widowed mother and sister to Birmingham, Alabama, where she built and managed an apartment building. By 1893 she was living in Fresno, California, where she ran a cattle ranch and vineyard until 1898.

THE FIRST WINDSHIELD WIPER

In the winter of 1902, while on a visit to New York City, she took a ride in a trolley car. It was a frosty day and she noticed that the trolley car driver had to drive with both his front windows open to the elements, otherwise the sleet that was falling would have obscured his view. She also noted how drivers had to stop their cars at the side of the road to wipe away the sleet and ice.

A solution to this inconvenient problem occurred to her and when she returned to Alabama where she was living again, she employed a designer to work on a hand-operated device that would keep a windshield clear. It consisted of a lever inside the vehicle's cab that controlled a rubber blade mounted on the outside of the vehicle. When the lever was operated, a spring-loaded arm would move back and forward across the glass. A counterweight ensured that the blade's rubber was in contact with the windshield as it swept across. She hired a local company to produce a working model and in 1903, was granted a 17-year patent for her windshield wiper.

DANGEROUS DISTRACTION

Mary attempted to make some money from her invention by trying to sell it to a Canadian company but, seeing no future in it, they replied to her: "we do not consider it to be of such commercial value as would warrant our undertaking its sale." In fact, many thought that the wiper would be a dangerous distraction for drivers and, anyway, many cars were built without windshield wipers at the time.

Her patent eventually expired in 1920 and as cars became increasingly popular, they began to enjoy the added bonus of windshield wipers along the lines of the one Mary Anderson invented. In 1922, wipers became standard equipment on Cadillacs.

It is interesting to know that the first automatic windshield wipers were also invented by a woman—Charlotte "Lotta" Bridgwood. Her Electric Storm Windshield Wiper used rollers instead of blades but, unfortunately, like Mary's invention, it was not a commercial success.

Windshield wipers are essential for a clear view of the way ahead.

VIRGINIA APGAR

TRANSFORMING THE FACE OF MEDICINE
INVENTION: THE APGAR SCORE

Pediatricians assess the health of newborn babies using the five criteria of the Apgar score: Appearance, Pulse, Grimace, Activity, and Respiration.

The American obstetrical anesthesiologist Virginia Apgar was born in Westfield, New Jersey, in 1909, into a family that she said "never sat down." She learned to play violin as a child and developed an early interest in science and medicine possibly as a result of her father being an amateur astronomer and inventor.

Deciding to pursue a career in medicine, she graduated from Mount Holyoke College in 1929, majoring in zoology. At college, she was remembered for her relentless energy. She played for seven sports teams, wrote for the college newspaper, acted in drama productions, and played violin in the college orchestra. You might have thought she would not have much time left for studies, but despite all these extra-curricular activities, her academic work remained exceptional.

THE PROSPECTS FOR WOMEN

Apgar began her medical training at Columbia University's College of Physicians and Surgeons where she was one of just nine women in a class of ninety. Completing her MD in 1933, she began a two-year surgical internship at Columbia-Presbyterian Hospital but when she was warned by her mentor that the prospects for women surgeons during the Depression did not look good, she decided that anesthesiology might be a safer bet.

After the second year of her internship she underwent a year's training as a nurse-anesthetist and undertook further studies at the University of Wisconsin and Bellevue Hospital in New York. She went back to Presbyterian as director of a new Division of Anesthesia that was being established in the Department of Surgery. For the next eleven years she worked tirelessly transforming the anesthesia service at Presbyterian into one staffed by physicians rather than nurses. She also became a highly respected teacher in the division's education program.

EVALUATING NEWBORNS

In 1949, Apgar was made a full Professor of Anesthesiology at Columbia University's College of Physicians and Surgeons, the first woman to hold that rank at the university. Around this time, she began to become interested in the effects of maternal anesthesia on newborn babies, seeking to lower neonatal mortality rates which remained disturbingly high.

By 1952, she had developed a scoring system that was designed to evaluate the health status of newborn babies. It was based on measurement of their heart rate, respiration, movement, irritability, and color at one minute and then five minutes after they had been born.

Her name was borrowed as an acronym for the five simple criteria—**A**ppearance, **P**ulse, **G**rimace, **A**ctivity, and **R**espiration. These five are marked on a scale from 0 to 2. These are then added up to give a score out of 10. A low score obviously indicates that a baby needs medical attention but does not necessarily indicate a long-term problem.

The Apgar evaluation became standard practice, and is now used to evaluate the health and well-being of newborn babies around the world. Virginia Apgar died in 1974 at age 65.

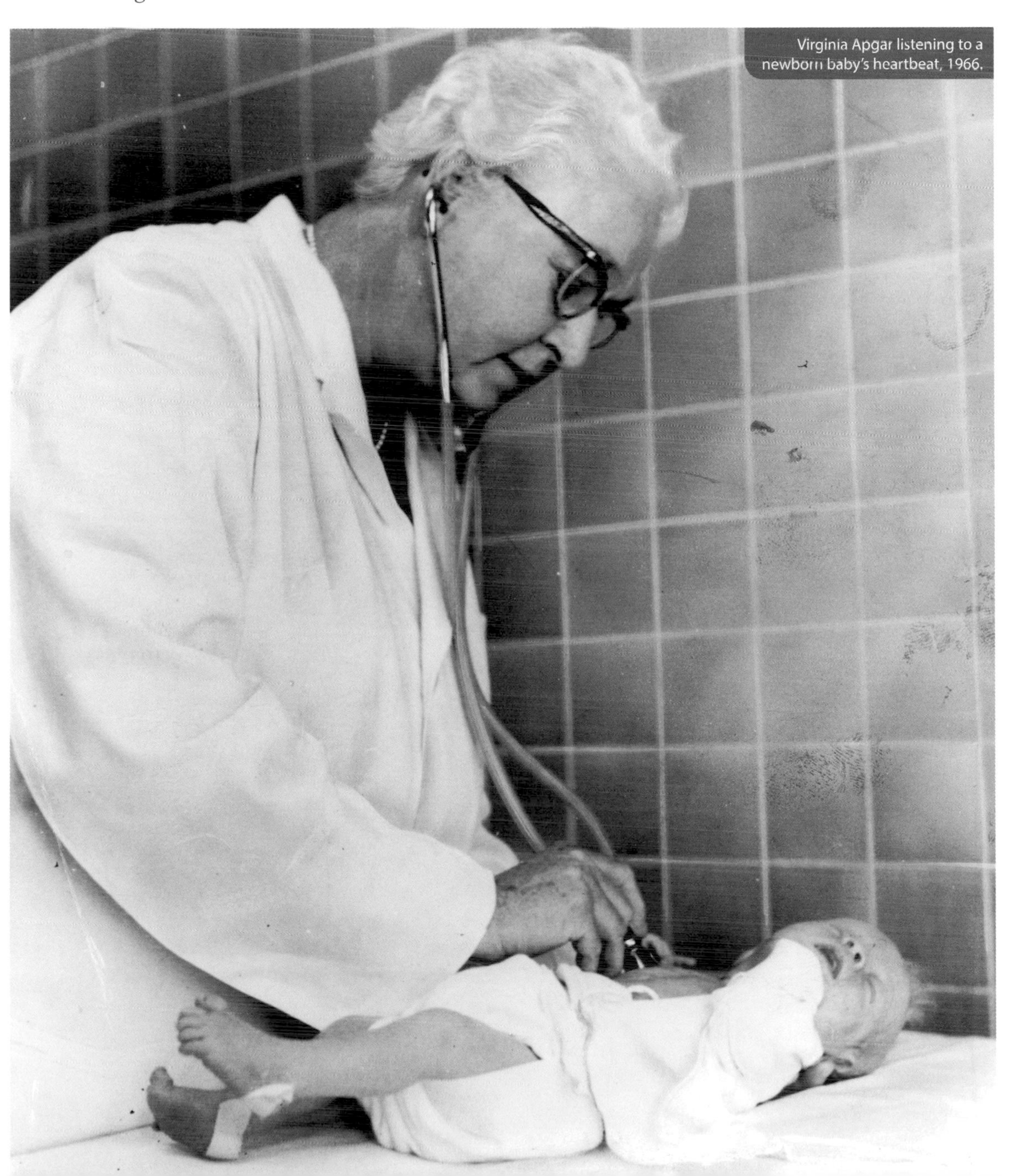

Virginia Apgar listening to a newborn baby's heartbeat, 1966.

ANCIENT WOMEN OF MEDICINE

In Ancient Egypt in 2700 BC, physicians freely mixed together the practices of religious ritual and medicine in an inventive scientific blend. The first ever woman medical practitioner is identified in a hieroglyphic carving near the step pyramid of Saqqara in the old capital city of Memphis. She was Merit-Ptah, the Chief Physician of the Pharaoh's court during the Second Dynasty. Hers is the first woman's name in history associated with medicine and science.

Octavia, sister of Roman Emperor Augustus, created many useful remedies such as her cure for toothache which used barley, flour, honey, and vinegar, mixed with salt, and baked before being pulverized with charcoal and scented with spikenard flowers.

Roman medical writer Galen mentions several women who were responsible for medical discoveries. The Greek Origenia invented remedies for diarrhea and spitting blood; Eugereasia created a remedy for inflammation of the kidneys; Xanita devised a cure for cold and fever; and a first century woman doctor, Antiochis, devised cures for diseases of the spleen as well as for arthritis and sciatica.

This mummy portrait of a young woman from Faiyum, Ancient Egypt, circa 120 – 130 AD, is exhibited in the Louvre, Paris. Portraits like this were painted on wooden boards to cover the faces of the mummified bodies of the dead. Due to the hot dry Egyptian climate, the paintings are frequently very well preserved, often retaining their brilliant original colors.

RUTH ARNON

SUPPRESSING MULTIPLE SCLEROSIS
INVENTION: COPAXONE

Kadeena Cox suffers from multiple sclerosis but won gold medals in both athletics and cycling at the Rio 2016 Paralympics.

The Israeli scientist and immunologist, Ruth Arnon, was born in Tel Aviv, Israel, in 1933. Taking her inspiration from her studious father Alexander who had degrees in electrical engineering and mathematics, Ruth resolved from an early age to study.

While in kindergarten she could already read and count and she went straight into second grade when she entered elementary school. Thanks to her mother's powers of persuasion, at the age of 13 she was enrolled at the Herzliya Hebrew Gymnasium, Tel Aviv, and by 15 she had decided that she wanted to work in the science of medicine.

She went to the Hebrew University of Jerusalem to study chemistry, spending her summers doing military training as many young Israelis did. After graduating in 1955 with an MSc degree, she served two years as an officer in the Israeli Defense Force. Around this time, she married an engineer named Uriel Arnon who was studying chemical engineering at the Technion, the Israel Institute of Technology in Haifa.

THE FIRST SYNTHETIC ANTIGEN

In 1957, she became a doctoral student at the Weizmann Institute of Science in Rehovot, Israel, where she was mentored by the Israeli immunologist Michael Sela. The two worked closely, publishing papers jointly in scientific journals and from 1975 to 1985 Sela was President of the Institute, while Ruth Arnon was Deputy President from 1988 to 1993.

It was while they were working together that they succeeded in synthesizing for the first time in the laboratory a substance that stimulated the body's immune system, creating the first synthetic antigen. An antigen is a molecule that is capable of inducing an immune response on the part of the host organism.

THE MIRACLE DRUG

With the help of another doctoral student, Dvora Teitelbaum, they discovered that their synthetic material could suppress multiple sclerosis in animals. The research and trialing of this material went on for the next thirty years until approval was granted for the resultant drug glatiramer acetate, also known by its trade name Copaxone. This miracle drug has become a major part of the fight against MS and is so effective that it helps sufferers lead an almost normal existence.

Copaxone was not Ruth Arnon's only success. She has also invented a synthetic influenza vaccine that is administered nasally. It will, it is expected, prove effective against a wide variety of flu virus for several years. In 2017 Professor Arnon was appointed co-chair of the UK-Israel Science Council, alongside Professor Lord Robert Winston.

DOROTHY ARZNER

LIGHTS, SOUND, ACTION!
INVENTION: THE BOOM MICROPHONE

A soundman holds a boom microphone above the heads of the actors on a film set as the camera and lighting rig rolls backward on tracks.

When she was young, a career in films never really entered the head of the inventor of the boom microphone, Hollywood director Dorothy Arzner (1897 – 1979). Indeed, she had harbored ambitions of becoming a doctor, and spent two years as a pre-med student and two years in the ambulance corps overseas. She had however, been close to the world of celebrities. She was raised in Los Angeles, where she worked as a waitress in her parents' restaurant frequented by movie stars such as Charlie Chaplin.

BLOOD AND SAND

Medicine lost out to the world of celluloid after Dorothy visited her first film studio, where she realized at once that she wanted to be a film director. She started in the film business as a typist for director William C. de Mille, at Famous Players-Lasky Corporation which later became Paramount Pictures, but was soon writing scripts and editing films. Her first editing job was on Rudolph Valentino's classic silent movie *Blood and Sand*, one of the biggest films of 1922.

After having worked on more than fifty films at Paramount, Arzner threatened in 1927 to move to the rival Columbia Studios unless she was allowed to direct her own movies. Paramount yielded and she was given *Fashions for Women*, a silent social comedy about a cigarette girl who falls in love with a count. A box office hit, it led to Arzner being offered more films to direct.

Film poster for *The Wild Party*, best known as Clara Bow's first talkie.

TALKING PICTURES ARRIVE

With the major success of the first talkie, Al Jolson's *The Jazz Singer* in 1927, silent movie stars were forced into the world of sound. Dorothy Arzner made history when she became the first woman to direct a sound picture, *Manhattan Cocktail* (1928), a mainly silent film which included some sound sequences.

One of the biggest names forced into the talkies was Paramount's top silent star Clara Bow. The personification of the 1920s sex symbol, "The It Girl," as she was known, was a huge box office draw. In 1929, Dorothy Arzner was given the onerous responsibility of directing the first film in which Clara Bow would speak, *The Wild Party.* It was also to be the first fully synchronized sound picture directed by a woman.

CLARA BOW'S STAGE FRIGHT

Things did not go well. Clara Bow had a thick Brooklyn accent that the silence of the pre-talkie era concealed. She was terrified of the transition to sound, and found it impossible to master the technique of talking into a microphone. The stressful situation was not helped by the sudden explosion of the microphone as she uttered her very first line.

Arzner came up with an ingenious solution to alleviate Clara's fear of sound. Working with her crew she rigged-up a microphone to a fishing rod which was dangled out of shot over the actors' heads, capturing Clara's less-than-lyrical spoken tones as she moved around the set. It was the first film studio use of the boom microphone, now an essential tool in filmmaking universally.

Sadly, Arzner failed to take out a patent on her invention, and a year later Edmund H. Hansen, a sound engineer at the Fox Film Corporation filed a patent for a very similar device. Although she did get some reward when *The Wild Party* became the third-highest grossing film of 1929.

GOLDEN AGE CAMERA WOMAN

More films followed, usually featuring free-spirited, aggressive, and independent women just like Dorothy Arzner, and included collaborations with Katharine Hepburn and Joan Crawford. In 1932, she left Paramount and began working independently.

Her career spanned from the silent era of the 1920s to the early 1940s. In her fifteen-year career as a director (1928 – 43), Arzner made three silent movies and fourteen "talkies." She was the only woman director during the "Golden Age" of Hollywood, the first woman to be invited to join the Directors Guild of America, and a role model for women filmmakers in future years.

Dorothy Arzner died in 1979, at age 82, in La Quinta, California. She was posthumously awarded a star on the Hollywood Walk of Fame in 1986.

Dorothy Arzner and Clara Bow on the set of *The Wild Party* (1929).

BARBARA ASKINS

MAKING THE INVISIBLE VISIBLE
INVENTION: AUTO-RADIOGRAPHIC IMAGE ENHANCEMENT

Working for NASA at the Marshall Space Flight Center in Madison County, Alabama, physical chemist Barbara Askins (born 1939) was asked to find a way of improving astronomical and geological photographs taken from space. They often turned out fuzzy and lacked clear definition which was a problem for scientists who needed to see the fine detail. Making the invisible visible is quite an achievement, but that is what Barbara Askins did. Her invention would have a greater impact than anyone could have imagined.

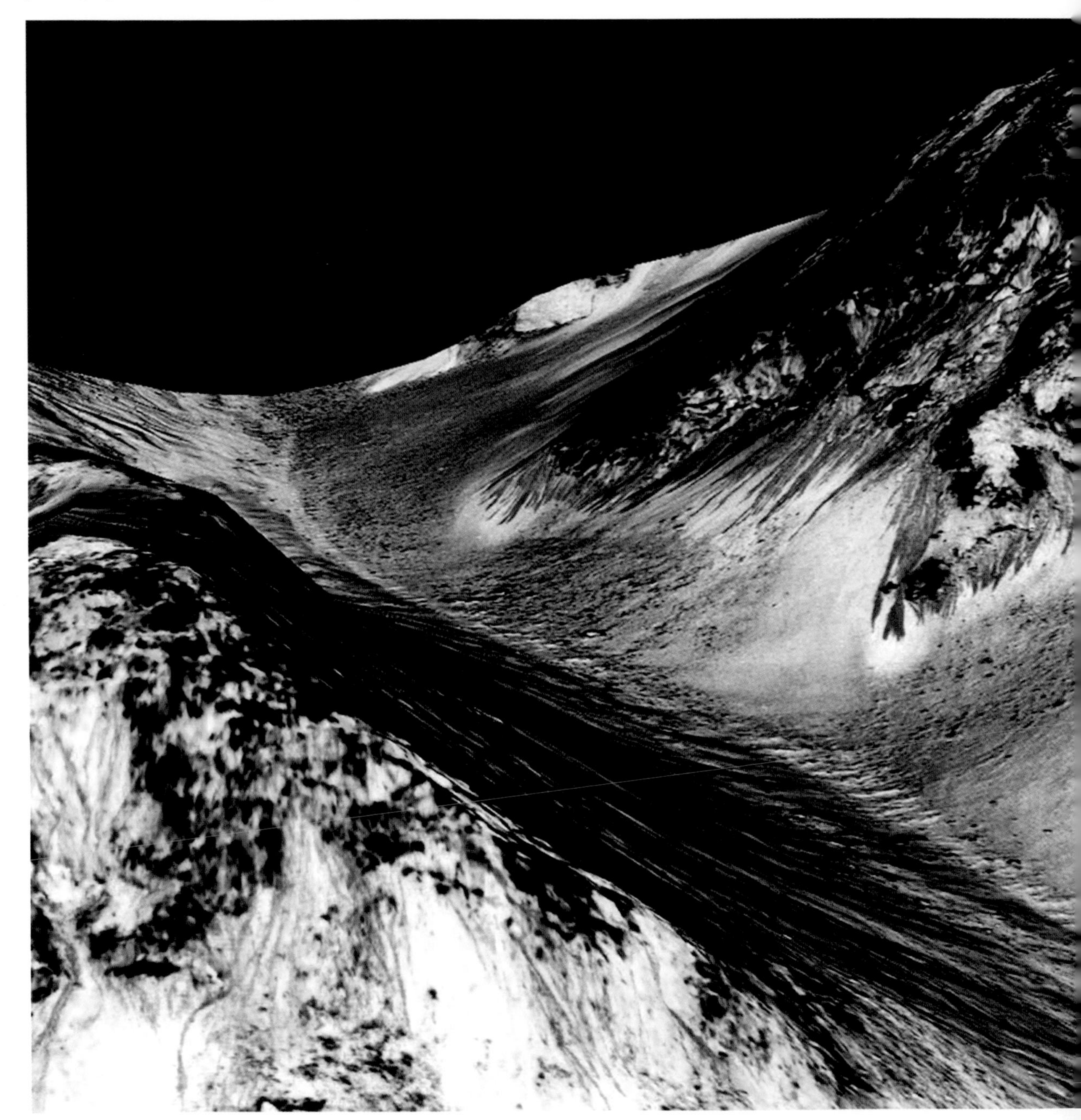

The process she developed was one in which "images on developed photographic emulsions can be significantly intensified by making the image silver radioactive and exposing a second emulsion to this radiation."

The print that resulted from this process was known as an auto-radiograph and it represented a huge increase in both density and contrast in the image. It could be used to tease data from underexposed images of space and was also useful in explaining the geology of other bodies in the solar system.

The process had a variety of other applications, including improvements in the clarity of X-ray images, allowing doctors to significantly reduce the amount of X-ray radiation used, and restoring old photographs. Askins patented her invention in 1978 and NASA employed it extensively for its research and development work.

For her invention of a new way of developing film, Barbara Askins was named National Inventor of the Year in 1978, the first women to win the award who had sole title to the patent for which she was being awarded.

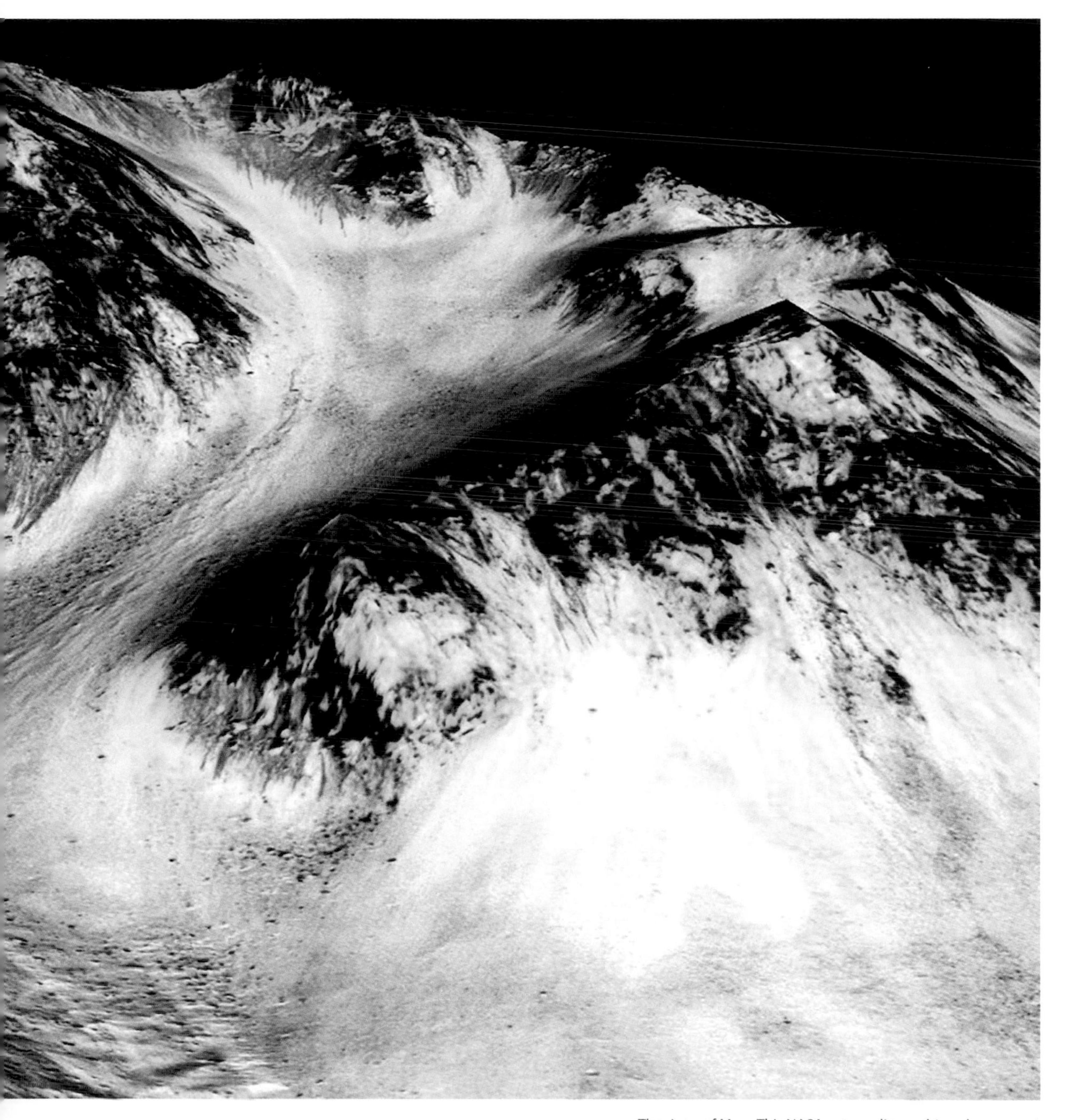

The rivers of Mars. This NASA auto-radiographic enhancement image provides the strongest evidence yet that liquid water flows intermittently on present-day Mars.

HERTHA AYRTON

THE SECRET OF RIPPLES
INVENTION: MATHEMATICAL LINE DIVIDERS

Sand ripples on the beach of Budle Bay, Northumberland, England.

Hertha Marks Ayrton was an award-winning English engineer, mathematician, inventor, and physicist, best known for her groundbreaking work on electric arc lamps and sand ripples. Born Phoebe Sarah Marks in Portsmouth, England, in 1854, she was the third child of an English seamstress and a Polish watchmaker who died in 1861, leaving the family in debt. In 1863, at the age of 9, she went to live with her aunt, the writer Marion Hartog, in north-west London, and attended the school she ran with her husband.

RADICALS AND EXILES

It was at Marion Hartog's school that Sarah discovered her love of science and mathematics. She mixed with the intellectuals, radicals, and exiles who were part of the Hartogs' social circle, and adopted the name "Hertha" given to her by friend Ottilie Blind, after the Germanic Earth goddess.

Ottilie, the daughter of German writer Karl Blind, was a well-connected young lady. She introduced Hertha to Madame Barbara Bodichon, a leading women's rights activist and early benefactor of Girton College for Women at Cambridge University. A university education for women was a new idea in the nineteenth century.

Madame Bodichon provided the Girton College funds for Hertha's studies and she passed the Cambridge University Examination for Women in 1874 with honors in English and mathematics.

MATHEMATICAL LINE DIVIDERS

At Girton, she organized the college's fire brigade and sang in the college choir. Hertha also launched her career as an inventor, devising a sphygmograph, a tool for recording the pulse in arteries.

In 1884 she patented a mathematical line divider, an instrument for dividing a line into any number of equal parts, which could be used by artists and architects. The device was unveiled at the Exhibition of Women's Industries and was her first major invention.

Hertha began to teach and also began a physics course in the evening at the Finsbury Technical College in London. Her teacher was Professor William Ayrton, an electrical engineer who had studied under Lord Kelvin at the Royal Society. He was also a widower with a young daughter, and Hertha and Ayrton fell in love and were married in 1885.

ELECTRIC ARC LAMPS AND SAND RIPPLES

Hertha Ayrton is celebrated for her work on the electric arc lamp, which was widely used at the time for street lighting. Their tendency to hiss and spark dangerously and produce a far from consistent light was a constant problem. Hertha proved that if air was excluded from the arc, the hissing disappeared and the light did not flicker. In 1895 she wrote a series of articles on her discovery for the *Electrician*, and in 1899, she was the first woman ever to read her own paper "The Hissing of the Electric Arc" to the Institution of Electrical Engineers (IEE). She was elected the first female member of the IEE shortly after.

By the late nineteenth century, her electrical engineering work was recognized internationally and she attended the 1899 International Congress of Women in London and spoke at the Paris International Electrical Congress in 1900.

In 1901 she started work analyzing the ripple patterns that appear in sand when a wave passes over it. This had been a scientific mystery until Ayrton read "The Origin and Growth of Ripple Marks" to the Royal Society in 1904. She was awarded the Royal Society's Hughes Medal for an original discovery in 1906, the first woman to do so and one of only two women to win it to date.

LIFE-LONG INVENTOR

Sadly, William Ayrton died in 1908, but Hertha continued inventing and in 1915 during World War I, she designed the Ayrton anti-gas fan to disperse poison gas from the British trenches. During the war, 104,000 of the Ayrton fans were issued to troops on the Western Front and they were in daily use from May 1916. She also invented improved searchlights that were used in both world wars.

Hertha fully supported women's rights, taking part in protest marches and allowing arrested suffragettes to stay at her house after their release from prison. She was also a life-long inventor and owns twenty-six patents: five on mathematical line dividers, thirteen on electric arc lamps, and the rest on air propulsion.

Hertha Ayrton died in 1923, at age 69. In 2010, she was named by the Royal Society as one of the ten most influential British women in the history of science.

THE ROYAL SOCIETY

The Royal Society initially grew out of meetings of groups of physicians and natural philosophers in the middle of the seventeenth century. In 1660, the formation of a "College for the Promoting of Physico-Mathematical Experimental Learning" was announced and in 1662, it was given a Royal Charter and renamed the Royal Society of London. It met at Gresham College. In the second half of the eighteenth century, it became customary for the government of the day to refer important scientific issues to the Society and its Fellows were appointed to serve on government committees relating to scientific matters.

By 1739, the membership had been increased to 300 and the Royal Society had flourished under the presidency of Sir Isaac Newton who occupied that position from 1703 to 1727. The Society's normal business continued to be the demonstration of experiments and the reading of important scientific papers.

In the nineteenth century, the Society entered a period of relative decline, reflecting, perhaps, a decline in science in Britain. Charles Babbage published *Reflections on the Decline of Science in England, and on Some of its Causes*, in which he was highly critical of the Royal Society. It was decided to limit membership to fifteen a year. This was increased to seventeen in 1930 and twenty in 1937. By 1941 membership was usually between 400 and 500. The first women members were elected in 1945.

SARAH TABITHA BABBITT

SHAKE IT ON DOWN
INVENTION: THE CIRCULAR SAW

An early eighteenth-century circular saw.

Sarah "Tabitha" Babbitt was born in 1779 in Hardwick, Massachusetts, and by 1793 was living in the Harvard Shaker Village working as a weaver. This particular community worked mainly in forestry, and every day Tabitha watched men working hard sawing logs. In 1810, she came up with a tool that would make their job a lot easier.

The men used a pit saw, a long two-handled saw which they pulled from side to side, one at each end. They cut along the length of the log to create planks. The problem was that the saw only cut into the wood on its forward pull, rendering the second pull only useful for taking it back into position. It was a waste of energy, time, and manpower.

Tabitha devised an alternative by making a prototype to demonstrate its effectiveness. It consisted of a circular blade attached to her spinning wheel. She used the pedal of the spinning wheel to power it and as the blade spun, no movement was wasted. Furthermore, it cut faster and the manpower required was reduced by one.

A larger version of her original design was installed at the local sawmill and soon other sawmills were copying her invention. But as a Shaker, Tabitha did not believe in patenting her idea and, therefore, made no money from it.

Tabitha Babbitt is also credited with inventing the spinning wheel head and a process for manufacturing the type of nails known as "cut nails." When she died in 1853, she was working on a new process for manufacturing false teeth.

THE SHAKERS

The United Society of Believers in Christ's Second Appearance, commonly known as the Shakers, is a Christian sect that was founded in England in the eighteenth century. The name "Shaker" derives from the description of them as "Shaking Quakers" because of their behavior during worship when they behave ecstatically.

Best-known today for its popular styles of nineteenth-century furniture and crafts, the Shaker legacy includes many achievements in social reform, agriculture, technology, and innovation. The flat broom, the circular saw blade, the spring clothespin, chair tilter buttons, and the paper seed envelope are all among a long list of Shaker inventions.

Shaker religious exercises in the meetinghouse, New Lebanon, New York.

BETTY LOU BAILEY

ENGINEER EXTRAORDINAIRE
INVENTION: THE VARIABLE EXHAUST NOZZLE

The Nimbus satellites (Nimbus 1 pictured here) made a lasting mark on meteorology and climate science.

Born in 1929, Betty Lou Bailey went into engineering at the suggestion of her sister Helen who had been taught how to weld by her husband. Women engineers were still very rare and Betty Lou was the only female in her graduating class of 700 engineers. She had entered the undergraduate program at the University of Illinois a year early and graduated with honors. Moving on to Penn State Graduate Center in King of Prussia, Pennsylvania, she graduated in 1967 with a Masters in Engineering.

From there, Betty Lou took a job with General Electric Company (GE) where she spent her entire career, holding various positions as a testing, design, and systems engineer. As her career developed, she moved from working on household appliances to steam turbines and jet engines. She eventually worked on the NASA Nimbus weather satellite project. Nimbus satellites were robotic spacecraft used for meteorological research.

GE had a gas turbine, the 7F, that was the first of its kind to be developed and made operational. One day, Betty Lou was working on the roof of the turbine when she spotted a leak from the gas pipe that led out of the gas valve compartment. She recognized immediately the critical nature of this leak. The company had counted all the gas coming from that compartment as going into the turbine, but the leak meant that not all of it was making it that far.

The leaked gas was mixing with the gas that was going back into the machine and carbon monoxide was destroying the seal. They had been running efficiency tests for six weeks and all of the information gleaned was rendered inaccurate by the leak. This would have led to the turbine failing the tests. She devised a means of stopping leakage—a convergent-divergent variable exhaust nozzle—and was granted the patent for it.

Betty Lou Bailey, engineer extraordinaire, died suddenly during a bicycle ride in 2007, at age 78.

ELLENE ALICE BAILEY

THE PROBLEM WITH POWDER PUFFS

INVENTION: THE POWDER PUFF

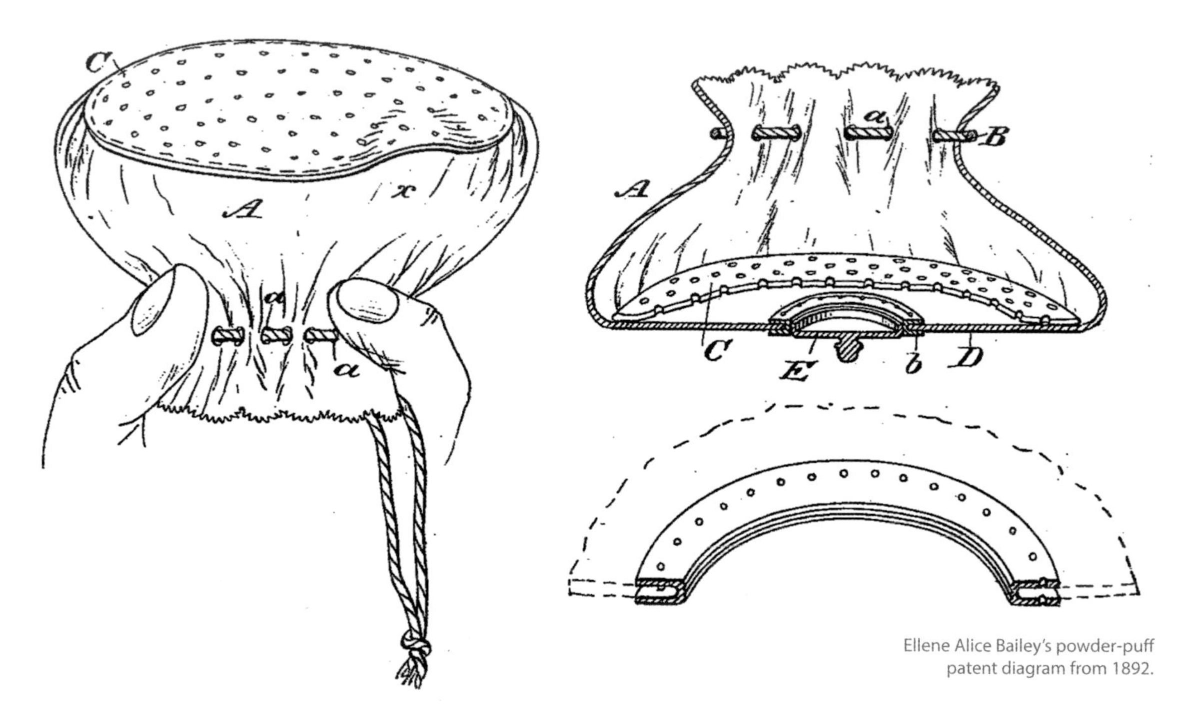

Ellene Alice Bailey's powder-puff patent diagram from 1892.

A newspaper article in October 1889 talked about a widow in New York City who earned a meager wage to support herself and her two young daughters as a piece-work embroiderer. A friend saw her make little embroidered bags containing a piece of cotton sprinkled with perfumed powder and a piece of swan's down. Legend has it that the friend liked what she saw and encouraged her to make more.

Although it is not entirely certain that this story is true, what is known for sure is that on July 3, 1882, Ellene Alice Bailey (1853 – 1897) filed an application for a patent on the powder puff at the US Patent and Trademark Office. The patent described her invention as "a powder receptacle or puff to be used for toilet purposes."

The problem with powder puffs was the control of the amount of powder released and preventing it from spilling and being wasted. With Bailey's powder puff, the bag that held the powder had a perforated flat bottom that allowed the user to control how much powder was being let out of the bag. It could also be carried conveniently and easily in a pocket.

Ellene followed it with the Thistledown powder puff, an invention from which she made some money as she sold a share of it. In 1889, she improved and simplified her first two puffs with the Floral powder puff. The Dainty followed in 1891. They were all financially successful.

Ellene was a prolific inventor. Her first patent in 1880 was named after her hometown. The Pond Fort boot was a high boot reaching to the knee that fitted tightly around the ankle. One of her most successful inventions was the Dart, a special needle for sewing on buttons, patented in 1884, 1886, and 1888. Ellene even threw herself into learning how to work the machinery and made the first 60,000 needles herself.

She also invented a device to ensure rubber overshoes stayed on, a silver whisk broom in 1887, a shaving case, a work box, a perforated felt chest protector, a sleeve holder, a manicure case, a wall album for photographs, a leg protector made of waterproof cloth, a corset shield, a detachable brassiere frame, chains for holding drapery, and many other items. This ingenious and industrious lady died in 1897, at age 44.

JANET EMERSON BASHEN

RECONFIGURING DATA
INVENTION: EQUAL EMPLOYMENT OPPORTUNITY SOFTWARE

Janet Emerson Bashen first came up with the idea of creating software to assist with Equal Employment Opportunity (EEO) in 2001 and was the first African American woman to hold a software patent. Bashen is CEO of Bashen Corp., a national labor and employment consulting firm that investigates EEO complaints and also sells the software to make the EEO complaint filing process easier. She has said: "I came up with the idea in 2001. Not everyone had a cell phone in 2001. I saw that papers in process got lost. There had to be a way to take in complaints—something Web-based and accessible away from the office."

Having borrowed $5,000 from her mother, her first customer was the restaurant chain, Denny's. She and the team she had assembled, worked for months on the design of the software and in 2006, after a rigorous vetting process by the Patent Office, she finally received the patent. By 2012, she was listed as one of the 100 most influential African Americans in entertainment, politics, sport, and business by *Ebony* magazine.

All employers must follow certain procedures for recruitment to ensure that all applications are treated appropriately and fairly and that, all things being equal, no candidate is rejected because of their age, gender, race, sexuality, or any other protected characteristic.

PATRICIA BATH

MAKING THE BLIND SEE
INVENTION: THE LASERPHACO PROBE

Dr. Patricia Bath with an anatomical model of an eye, 2012.

Patricia Bath is no stranger to being first. Born in Harlem, New York, in 1942, she was the first African American to complete a residency in ophthalmology at New York University and the first African American woman doctor to be granted a medical patent. She is also the founder of the American Institute for the Prevention of Blindness in Washington DC. This institute describes eyesight as "a basic human right."

From an early age she was always encouraged to pursue academic interests by her parents. When she was 16 she attended a cancer research workshop that was run by the National Science Foundation during which she impressed the program head, Dr. Robert Bernard, who delivered her findings at a later conference.

It took her only two years to graduate from high school following which she enrolled at Hunter College, Manhattan, and was awarded a bachelor's degree in 1964. She graduated with honors in medicine from Howard University, Washington DC.

In 1968, she became an intern at Harlem Hospital while pursuing a fellowship in ophthalmology at Columbia University, and discovered that African Americans were more likely than her other patients to suffer from blindness and glaucoma. This led to her developing a community ophthalmology system to help those unable to afford eye treatment.

In 1974, she became Assistant Professor of Surgery at both Charles R. Drew University and the University of California. The following year she became the first woman faculty member in the Department of Ophthalmology at UCLA's Jules Stein Eye Institute.

In 1981 she began working on her remarkable invention, the Laserphaco Probe. It used laser technology to provide a less painful and more precise treatment for cataracts. The patent for the device was granted in 1988. The laser dissolves the cataract painlessly before irrigating and cleaning the eye and inserting a new lens. This miraculous tool helps restore the eyesight of patients who had been blind for up to thirty years. In 2000 she patented a method of cataract treatment using ultrasound.

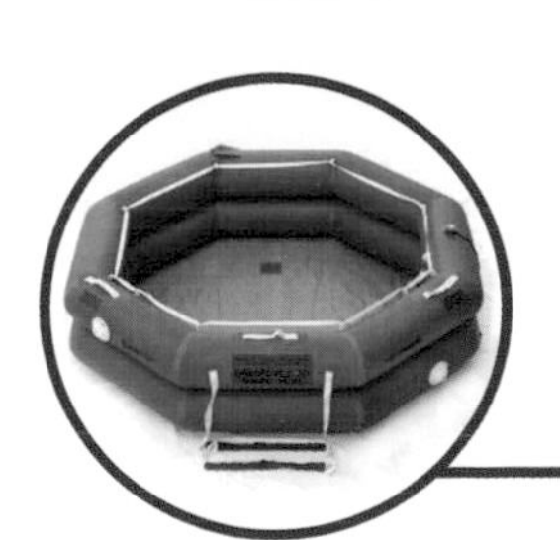

MARIA E. BEASLEY

FOR THOSE IN PERIL ON THE SEA
INVENTION: THE LIFE RAFT

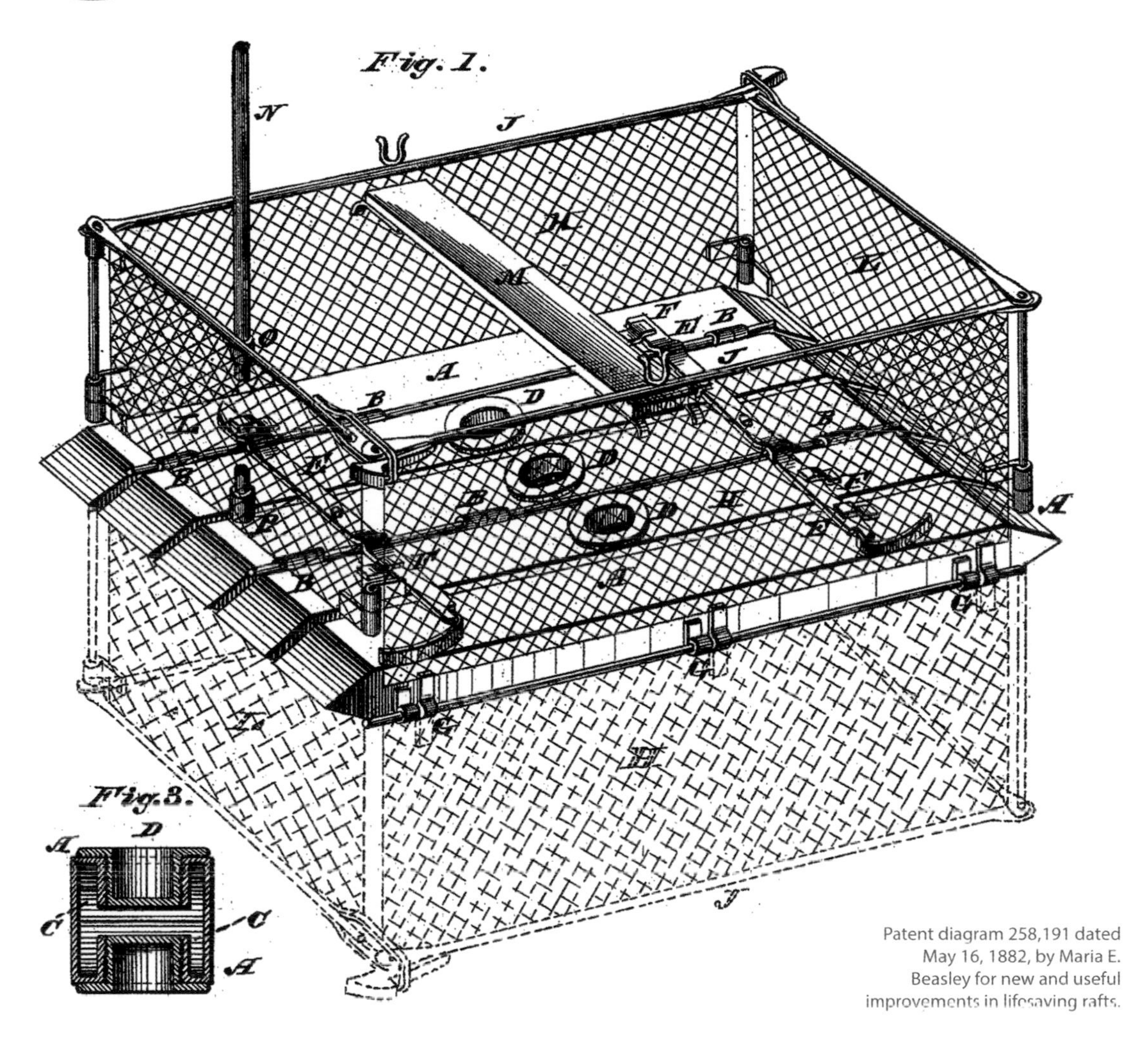

Patent diagram 258,191 dated May 16, 1882, by Maria E. Beasley for new and useful improvements in lifesaving rafts.

Maria E. Beasley was an inventing machine. Born in Philadelphia in 1847, she held fourteen patents in the United States and Canada, her first granted in 1878, for a barrel-hooping machine. She demonstrated it at the World Industrial and Cotton Exposition in Jackson Park, Chicago. It made her rich, the Washington newspaper, the *Evening Star*, noting in 1889 that she "made a small fortune out of a machine for the manufacture of barrels." The machine could make 1,500 barrels a day.

At the same exposition, she also demonstrated her improved design for a life raft, one that was "fire-proof, compact, safe, and readily launched." Patented in 1882, her life raft had guard rails and rectangular metal floats while life rafts of the time had hollow tube floats and no railings. She designed it so that it folded and unfolded and could be stored easily. Her life rafts were used on the *Titanic*.

All that we really know about Maria Beasley is that in 1880 she was recorded in the census as a housewife. But she was rapidly becoming a successful and innovative inventor and entrepreneur, and between 1891 and 1896, Chicago directories listed her occupation as inventor. Some of the other inventions she patented were foot warmers, cooking pans, and anti-derailment devices for trains. She was soon earning almost $20,000 a year, a small fortune at the time.

RUTH BENERITO

SAVING THE COTTON INDUSTRY
INVENTION: WASH-AND-WEAR COTTON FABRICS

Ruth Benerito working in the USDA's Chemical Cotton Reactions Laboratory.

How many hours of our lives are wasted standing at an ironing board, pressing the wrinkles out of clothing? Ruth Benerito (1916 – 2013) must have truly hated it because she devoted her life to removing the ironing board from our lives.

As is the case with many inventors, Benerito benefited from the encouragement of enlightened parents. Her father, John, a civil engineer, was a champion of women's rights and her mother, Bernadette, was an artist, described by her daughter as a "truly liberated woman." Ruth was born in 1916 in New Orleans. At 15, she enrolled at H. Sophie Newcomb College before going on to the women's college at Tulane University where she gained a Bachelor of Chemistry degree in 1935. A master's followed and she completed her studies with a PhD in chemistry from the University of Chicago.

In 1953 she got a job at the United States Department of Agriculture's Southern Regional Research Center in New Orleans, and spent most of her working life there. She started out in the Intravenous Fat Program of the Oil Seed Laboratory and two years later she was appointed project leader. At this time, she helped to develop a fat emulsion for feeding soldiers wounded in the Korean War intravenously. It was also used in hospitals.

In 1958, she was promoted to acting head of the Colloid Cotton Chemical Laboratory and a year later became research leader of the Physical Chemistry Research Group of the Cotton Reaction Laboratory. Later, she also taught, as adjunct professor at Tulane University and lectured at the University of New Orleans.

Her most famous work is related to the use of monobasic acid chlorides in the production of cotton, and she has been granted no fewer than fifty-five patents in this area. It results in wrinkle-free, more durable garments, wash-and-wear as they are popularly described. Her work also paved the way for stain- and flame-resistant fabrics and was described by the Chemical Heritage Foundation as having "saved the cotton industry."

MIRIAM BENJAMIN

PUSHING THE BUTTON!
INVENTION: THE GONG AND SIGNAL CHAIR FOR HOTELS

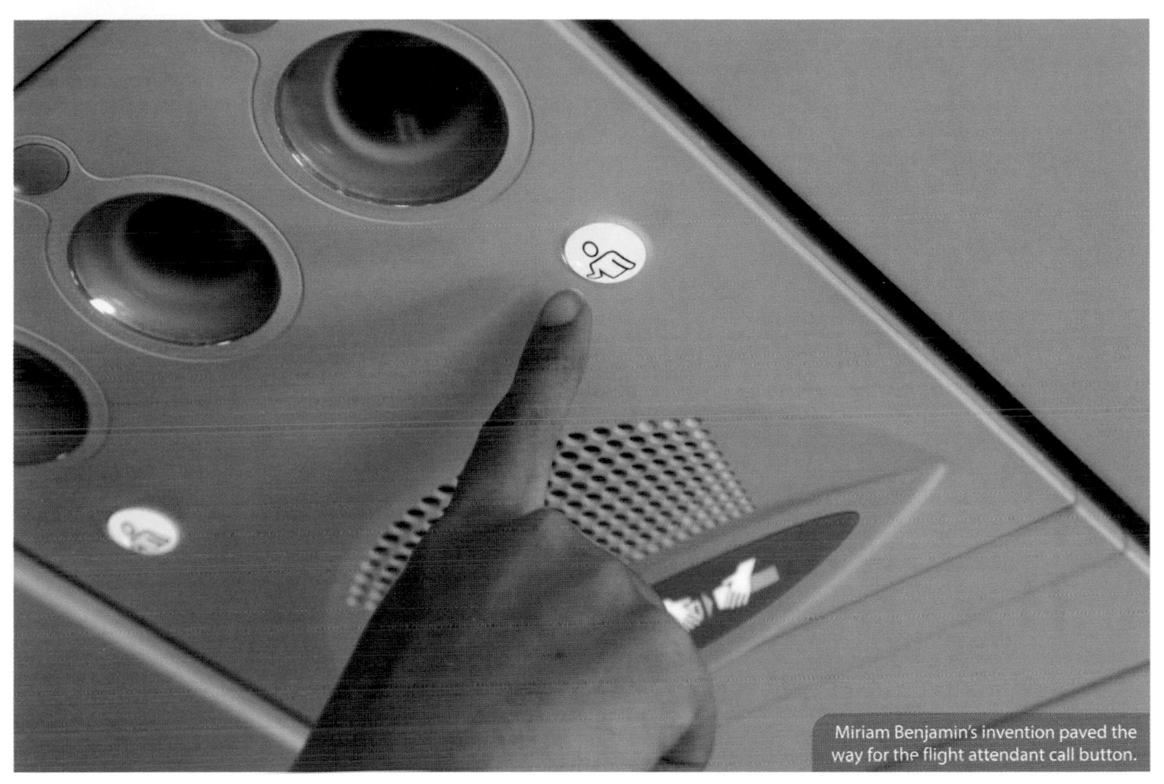

Miriam Benjamin's invention paved the way for the flight attendant call button.

Little did Miriam Benjamin realize her invention, or at least a version of it, would one day be used to summon a gin and tonic in an airliner seven miles above the surface of the earth.

Benjamin, born in Washington DC in 1861, was an African American teacher and inventor who, in 1888, obtained a patent for an extraordinary invention—the Gong and Signal Chair for Hotels. The idea was that a hotel guest seated in the chair would press a small button on the back of it that would send a signal to a waiting hotel employee. The pressing of the button would also illuminate a light on the chair, enabling the guest requiring attention to be immediately identified. It was the precursor of the airline flight attendant call button.

Miriam Benjamin was only the second African American woman to be granted a patent. She died in 1947, at age 86.

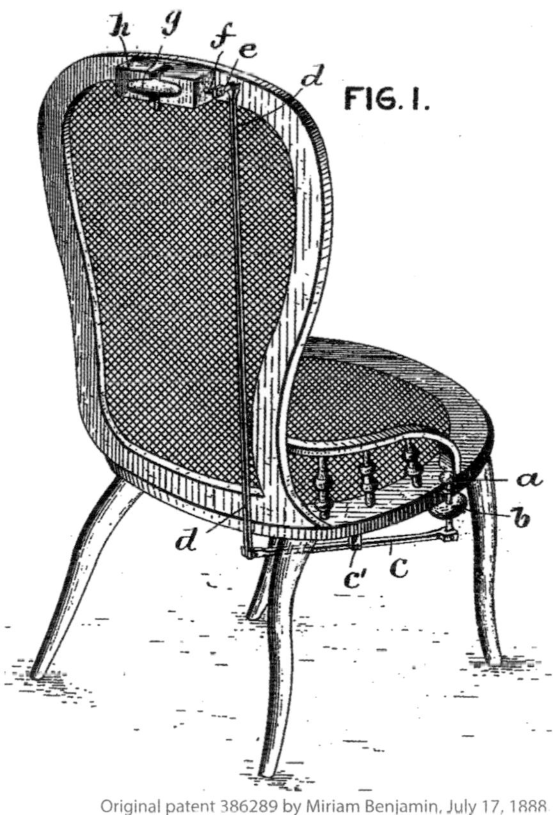

Original patent 386289 by Miriam Benjamin, July 17, 1888

MELITTA BENTZ

SO SIMPLE BUT BRILLIANT!

INVENTION: THE COFFEE FILTER

Melitta Bentz started her coffee filter business in 1908 and the company continues to enjoy great success today.

Coffee geeks can find Melitta coffee filters on the shelves of every supermarket today and many variations are also available. Their inventor was a German housewife who quite simply got tired of brewing the coffee too much, resulting in a grainy, bitter beverage. By 1908, Melitta Bentz, who was born in Dresden, Germany, in 1873, was experimenting, trying to make a smoother coffee and using anything to hand.

Eventually, she decided to use the blotting paper in her son's school exercise book in conjunction with a brass pot that she had perforated with a nail. It was so simple, but it was brilliant! Everyone loved her coffee, free of coffee grounds and smooth, having none of the previous tongue-clenching bitterness. She applied for a patent and it was granted on June 20, 1908, for a "Filter Top Device Lined with Filter Paper." She immediately went into business with the princely sum of 73 pfennig. With a local tinsmith engaged to manufacture the devices she had designed, she took a stand at the 1909 Leipzig Fair and immediately sold 1,200 coffee filters.

The business went from strength to strength and in 1911 was awarded a gold medal at the International Hygiene Exhibition in Dresden. World War I disrupted things but after the war the re-established company began to grow, as did demand for the product. By 1928 there were 80 employees. In the 1930s, the filter top was tapered so that it formed the shape of a cone. This gave it a larger filtration area that was lined with ribs, providing improved extraction of the ground coffee.

Melitta passed the business on to her sons, Horst and Willi, but still ensured that the workers were well taken care of, with paid holidays and a five-day working week. After World War II, production resumed in 1948 and it has never looked back. Melitta passed away in 1950 but Melitta Group KG is now run by her grandsons, Thomas and Stephen Bentz, employing 3,300 employees.

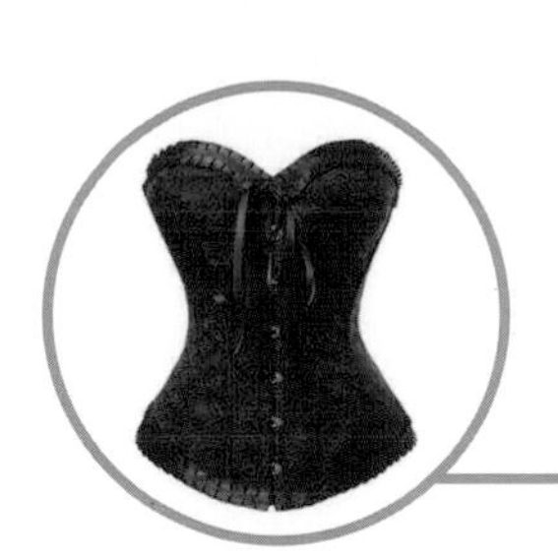

MADAME ROSA BINNER

THE CORSET QUEEN

INVENTION: A LIGHTWEIGHT CORSET

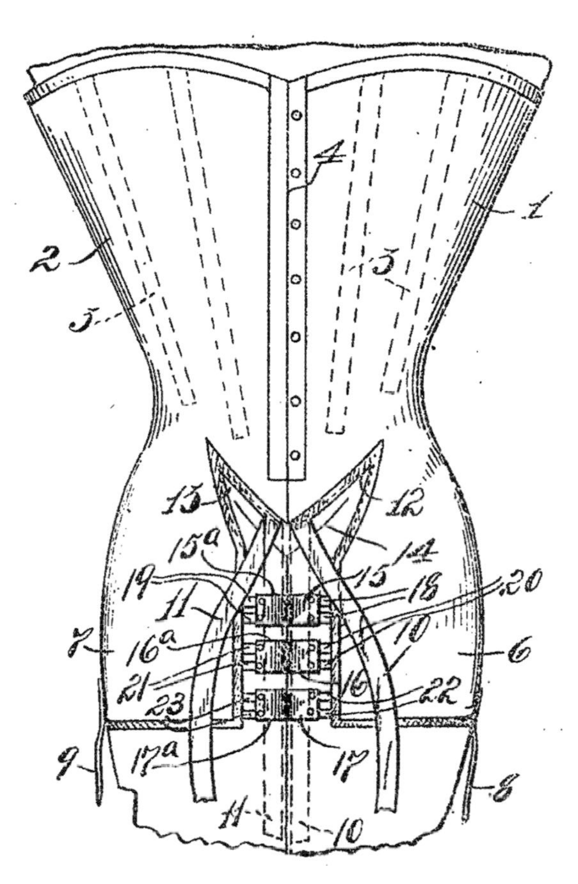

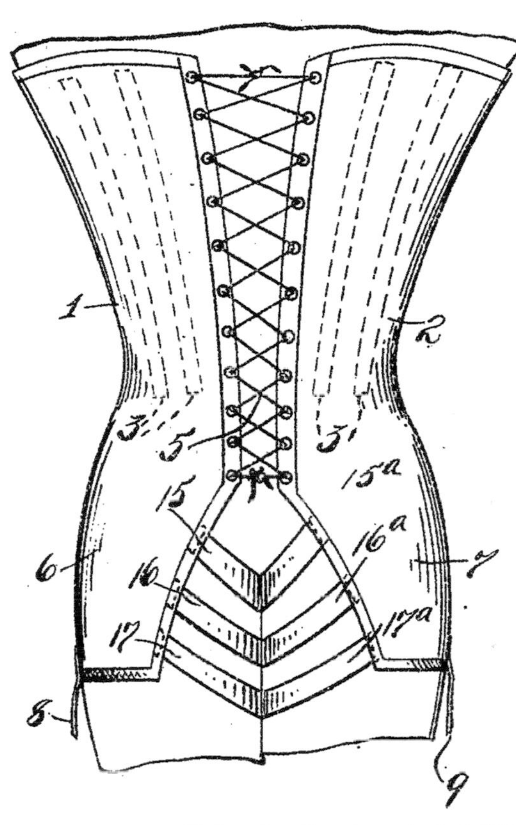

Madame Rosa Binner's corset patent of 1909.

In the 1920s, young women began to go off the idea of figure hugging and rib squeezing corsets, which marked the end of the road for many corset designers and manufacturers. Not the Vienna-born Madame Rosa Binner, however. She had been making corsets since the age of 11 and worked for a corset maker in Europe. She emigrated to the United States in 1896, but as she spoke no English she struggled to even realize she was living in Hoboken, New Jersey, which she thought was New York City. Eventually she settled in a New York apartment and began to make corsets for maids. The wealthy employers of the maids were soon ordering corsets too and her little business grew.

When she first arrived in America, she was shocked by the fashion for the hourglass figure that she insisted caused damage to women's bodies. When x-rays became available in America, Madame Binner sent customers to doctors so that they could be shown the damage that was being done to their ribs by the fashion of the time. By way of thanks, doctors would prescribe that patients go to her for the lighter, more sensible corsets she designed and sold. "Now the girls are flying airplanes," she said. "They must adapt their corsets to the air. Their corsets must expand as the flyers change positions—or make parachute jumps."

She famously designed a gartered corset for the celebrated American actress and singer Lillian Russell that had diamond buckles and cost $3,900. It was reported that, when Russell's house caught fire, she screamed at firefighters: "Never mind my jewels. Rescue my corset!" Inevitably, Hollywood sought her out and she designed corsets for many of Twentieth Century Fox's young starlets. Judy Garland, Mae West, and Mrs. John D. Rockefeller were among her other clients.

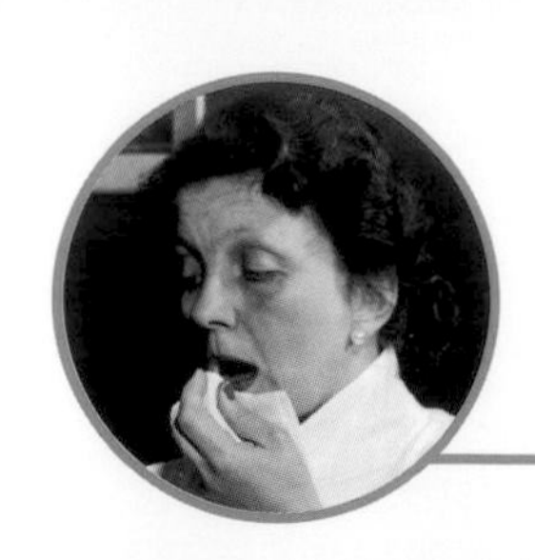

HAZEL BISHOP

SEALED WITH A KISS

INVENTION: THE FIRST LONG-LASTING LIPSTICK

Hazel Bishop's No-Smear Lipstick advertisements from the 1950s.

Hazel Bishop was born in Hoboken, New Jersey, in 1906, the child of a successful businessman. She had originally intended to be a doctor but her studies were interrupted by the Wall Street Crash of 1929. By 1935, Hazel was employed as a research assistant and in 1942, she began working as an organic chemist for Standard Oil Company where she designed fuels for aircraft fighting in World War II.

In her spare time, she began developing ideas, and invented a lipstick that would not smear or smudge and that was long-lasting. It was tough work keeping your lipstick tidy all day in the 1940s, especially when kissing! Experimenting with staining dyes, oils, and molten wax in her home kitchen, Hazel used her biochemistry experience and came up with a product that she named straightforwardly, "No-Smear Lipstick."

In 1950, she founded the cosmetics company Hazel Bishop Inc. to manufacture and market her lipsticks and started selling in retail outlets for $1 a stick. "Never again need you be embarrassed by smearing friends, children, relatives, husband, sweetheart," the early advertising said. It was guaranteed kiss-proof with the catchphrase "stays on you … not on him."

Her lipsticks were snapped up and were sold out by the end of the first day on sale. Soon Bishop was featuring on the cover of *Business Week*, the first woman to do so. Her success led to what was known as the "Lipstick Wars" as other cosmetic manufacturers such as Revlon brought out their own versions of No-Smear Lipstick.

Despite the competition, Hazel Bishop Inc. continued to do well, capturing more than 25 percent of the lipstick market. Three years after launching, her sales revenue was astonishingly more than $10 million. But in 1954, she lost control of her company to majority shareholders led by Raymond Spector. Hazel Bishop was thrown out and banned from using her own name in other business ventures.

She did not retire, however. Instead, she founded a research laboratory and worked for the National Association of Leather Glove Manufacturers, developing Leather Lav, a leather glove cleaner. She also developed a foot-care product and in the late 1950s she invented a solid perfume stick, Perfemme. In 1978, she was made a professor at the Fashion Institute of Technology in New York. Her remarkable life ended in 1998 at the age of 92.

SARA BLAKELY

KEEPING YOUR BUTT COVERED

INVENTION: SPANX UNDERWEAR

Sara Blakely displaying her SPANX body-shaping underwear.

The Florida heat made door-to-door selling in a formal outfit an uncomfortable job for Sara Blakely. She hated the way her pantyhose toes looked when she was wearing open-toed shoes. The rest of the pantyhose did a great job of eliminating VPL very well. It was always the toes. One night when she was getting ready for a party, Sara realized she didn't have the right underwear to provide a smooth look under her white pants. Armed with scissors, she cut the feet off her control-top pantyhose and her brilliant idea was born!

At age 27, Sara spent the next two years and her $5,000 savings on developing the concept, and wrote her own patent. She drove to North Carolina to sell her new product to the hosiery manufacturers located there. It seemed like no one was interested, and she went home. But a couple of weeks later, however, Sara got a call from one of the manufacturers who had had second thoughts. One of his daughters had persuaded him that the idea had potential. He agreed to help produce a prototype.

Working on feedback from her friends and family, Sara adjusted the prototype and introduced another innovation. Until then hosiery manufacturers had worked on a one-size-fits-all basis, but Sara's underwear introduced different sized waistbands with inbuilt support and shaping. She called her product and her company "SPANX" and registered it as a trademark.

She persuaded Neiman Marcus to stock her underwear in seven stores and other major retailers soon followed. In November 2000, Oprah Winfrey named SPANX on her hugely popular annual list of "Favorite Things." The first year's sales were $4 million and rose to $10 million in the second year. Featured on QVC shopping channel SPANX sold 8,000 pairs in six minutes.

Since then the SPANX brand has grown to offer bras, underwear, leggings, arm tops, bodysuits, maternity wear, and even physique-improving underwear for men. Sara Blakely, now age 46, was named the ninetieth most powerful woman in the world by *Forbes* magazine in 2016, with a net worth of $1.12 billion.

HELEN BLANCHARD

THE ZIG-ZAG STITCH SEWING MACHINE
INVENTION: SEWING MACHINE INNOVATIONS

One of Helen Blanchard's original zig-zag buttonhole sewing machines as seen at the Smithsonian's National Museum of American History.

In common with many of the female inventors of the nineteenth century Helen Blanchard (1840 – 1922) had no training whatsoever. She did not receive her first patent until she was in her thirties, although she was undoubtedly intellectually agile and inventive from an early age. Born in Portland, Maine, into a wealthy family, she was one of six children. But her ship-owning father was ruined in the financial crash of 1866, leaving his family practically destitute.

It was at that point that Helen moved to Boston, Massachusetts, and began to develop her inventions, borrowing the money needed to obtain her first patent in 1873. This was for a sewing machine which could sew a zig-zag stitch that when applied to a seam gave a garment extra strength. By varying the depth of the needle, the stitch could itself be varied. Her original machine is now to be found on display at the Smithsonian's National Museum of American History.

Numerous other sewing innovations followed. She invented a method of producing a strong elastic stitch; a means of strengthening elastic goring for shoes; an improved method for welted and covered seams; a spool case that protected the spool from getting dirty, damaged or unwound; a new sewing needle that could be threaded with one hand; a machine for sewing hats; a surgical needle with a lancet point that was able to pierce the skin more easily, causing less discomfort and pain; and many others.

In all, she was granted twenty-eight patents over a period of 45 years, twenty-two of which were associated with sewing and sewing machines. She became a wealthy woman and was proud in 1901 to be able to buy back the property that her family had lost in the financial crash. In 2006, she was admitted into the National Inventors Hall of Fame.

KATHARINE BURR BLODGETT

THE INVISIBLE INVENTOR
INVENTION: LOW-REFLECTANCE "INVISIBLE" GLASS

Physicist Katharine Burr Blodgett at the General Electric Research Laboratories, 1938.

Born in 1898 in Schenectady, New York, Katharine Burr Blodgett was a brilliant student, earning a science scholarship to the prestigious Bryn Mawr College, Pennsylvania, at only 15 years of age. She was the first woman to be awarded a PhD in physics from the University of Cambridge in 1926, and was hired by General Electric as a research scientist straight after finishing her master's degree. There she worked alongside an old friend of her father, the eminent chemist Irving Langmuir. Blodgett was the first woman chemist to be employed by General Electric.

Langmuir's research into surface chemistry looked at how substances adhere to each other at the molecular level. He had established a means of creating thin films just a molecule thick on the surface of water. It was a process they extended to lipids, polymers, and proteins. Blodgett carried out research into applications for the films that played a big part in Langmuir winning the Nobel Prize for Chemistry in 1932.

Blodgett did not stop there, however. She worked out how to add layers to the microscopically thin films. This was done by dipping a metal plate into water that was covered by a layer of oil. When she dipped it a number of times, the result was layers of oil being stacked on the plate. She devised a means of controlling the thickness—to within a molecule—of the films she was creating, and using this technique she invented "invisible" glass. She added layers of film to each side of a sheet of glass until the visible light reflected by the layers cancelled out that reflected by the piece of glass. This type of non-reflective coating is now called Langmuir-Blodgett film.

The blockbuster movie *Gone with the Wind*, released in 1939, owes its crystal clear cinematography to Blodgett's work. Movies produced using non-reflective lenses for cameras and projectors continued to stun audiences into the period after World War II.

There were also uses during the war, with Blodgett's glass being used in spy cameras on aircraft as well as in the periscopes of submarines. Today her invention is used in computer screens, spectacles, windscreens, in fact anywhere in need of a perfectly transparent surface. She was issued with eight US patents during her career and was the sole inventor on all but two. She was also the inventor of poison gas adsorbents, methods for de-icing aircraft wings, and improving smokescreens. Katharine Burr Blodgett worked for General Electric until 1963. She died in 1979, at age 81, leaving behind an extraordinary legacy. In 2007 she was inducted into the National Inventors Hall of Fame.

MARIE-ANNE BOIVIN

THE ART OF OBSTETRICS

INVENTION: THE PELVIMETER AND VAGINAL SPECULUM

The vaginal speculum is used as a medical tool to hold open the vagina during medical investigation.

French midwife, writer on obstetrics, and inventor, Marie-Anne-Victoire Gillain (1773 – 1841) was born at Versailles near Paris, France, and educated at a convent in Étampes. During the French Revolution, the convent was destroyed but Marie-Anne went on to study midwifery and anatomy for three years. In 1797, she married a government official, Louis Boivin, but he died not long after their marriage, leaving her to bring up their daughter on her own.

She found work as the superintendent of a local hospital, opening a school of obstetrics during her time in that role. Still furthering her medical studies, she became a student and assistant of Marie-Louise Lachapelle, head of obstetrics at the Hôtel Dieu, Paris's oldest hospital. In 1800 she gained her diploma and began practicing at Versailles.

Madame Boivin's inventions included a pelvimeter and a vaginal speculum. The former was to more easily measure the diameter and capacity of a woman's pelvis, while the latter was used to dilate the vagina of an expectant mother and permit the examination of her cervix.

Marie-Anne Boivin is thought to have been one of the first medical practitioners to use a stethoscope to listen to the heartbeat of a fetus. She was also an innovative and skilful gynecological surgeon and contributed many other discoveries to the science of obstetrics, discovering the cause of certain types of bleeding and the causes of miscarriages. She also did a great deal of research into diseases of the placenta and the uterus. The medical textbooks that she wrote were translated into different languages and used for 150 years. Mme Boivin has been called one of the most important women in medicine in the nineteenth century. Her declining years sadly brought misfortune and unhappiness. She suffered bankruptcy and died of apoplexy in 1841 at the age of 68.

VANNA BONTA

SEX IN SPACE

INVENTION: THE 2SUIT SPACE SUIT

Vanna Bonta with her zero gravity "2suit" which was said to be one small step toward humankind colonizing the universe.

How to have sex in space is a question that is often in people's minds, but rarely on their lips. Fortunately, actress, writer, and inventor, Vanna Bonta was not afraid to explore the question. In 2008, she came up with the 2suit, a garment designed to enable intimacy between two people in the zero gravity environment of outer space. The aim, she said, was to allow humans to colonize planets beyond Earth and perpetuate the species. A prototype was sponsored by the History Channel and she and her husband donned the suit to test it in simulated zero gravity conditions. They demonstrated its efficacy by kissing while they were wearing it. Regrettably, this fun woman died in 2014, at age 56.

SARAH BOONE

IRONING OUT THE WRINKLES
INVENTION: THE IRONING BOARD

The ironing board is one of those things you just think must always have been there, at least since the invention of the iron. Not true. In fact, up until the late nineteenth century, people generally balanced a piece of wood between two chairs or tables, placed a blanket on it and pressed their clothes on that. Sarah Boone changed all that.

She was a rarity in the nineteenth century—an African American female inventor. Born Sarah Marshall in North Carolina in 1832, she moved to Connecticut before the start of the American Civil War where in 1847 she married a freedman named John Boone with whom she had eight children. She was working as a dressmaker at the time.

On April 26, 1892, Sarah was granted the patent for her improvements to the ironing board. Boone's ironing board was made of a narrow wooden board, with collapsible legs and a padded cover and was specifically designed for the fitted clothing worn during that time period.

The narrow, curved, wooden board was shaped in a way that permitted the user, as the patent application said, "to produce a cheap, simple, convenient and highly effective device, particularly adapted to be used in ironing the sleeves and bodies of ladies' garments." There had been other ironing boards before, but the crucial difference with Sarah Boone's ironing board was that it had a narrow end for ironing sleeves.

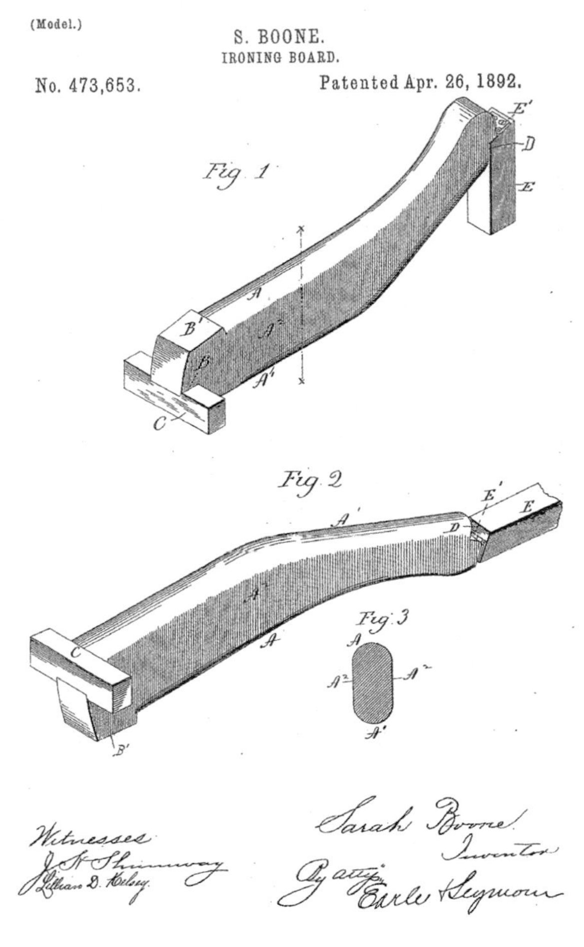

Sarah Boone's ironing board patent was particularly adapted for ironing sleeves.

IRONING BEFORE SARAH BOONE

No one knows exactly when we started to feel the need to iron our clothes, although the Chinese are said to have used metal pans filled with hot coals to take the wrinkles out of garments more than a thousand years ago. Real *iron* irons started being made in the Middle Ages when irons weighing between five and nine pounds were being made by blacksmiths. These were called "sad" irons, ("sad" being an Old English word for "solid") or "flat" irons, shaped pieces of metal that were flat and polished on the bottom and had a handle attached to the top. These irons were often made in the shape of a triangle enabling the user to iron the fabric around buttons or fasteners. The notion of having a wooden handle came, of course, from a woman, a Mrs. Mary Florence Potts who added a detachable walnut handle to her iron in 1871.

It was the French who invented electric irons, in 1880, but it was not until 1903 that they really took off. Earl Richardson, owner of the Pacific Electric Heating Company, based in California introduced the Hotpoint iron, the hottest point being at the front and not the center of the device. It was smaller and lighter than previous irons. The Hotpoint name went on to become famous in the manufacture and sale of electrical appliances.

YVONNE BRILL

A PIONEERING SPIRIT IN SPACE
INVENTION: PROPULSION SYSTEM FOR SATELLITES

Yvonne Brill was presented with the National Medal of Technology and Innovation by President Obama in 2011.

To be a woman rocket scientist in the 1940s was unusual to say the least. In fact, Yvonne Brill (1924 – 2013) was the only woman engaged in that particular discipline at the time. Naturally, such a distinction brought obstacles and problems. At the University of Manitoba where she wanted to study engineering, she was refused entry because there was no accommodation for women at an outdoor engineering camp which students were required to attend. She was forced, therefore, to study chemistry and mathematics. There was also the matter of having three children to look after, no mean feat when you have a career such as hers.

Nonetheless, in 1967 she still found time to invent a propulsion system that helped to keep communications satellites from moving out of their orbits. The hydrazine resistojet enabled satellites to carry less fuel and a bigger payload. It also gave better performance and greater reliability. They could also stay in space longer. It became the standard for the industry and earned her many accolades, including the National Medal of Technology and Innovation which she was presented by President Obama in 2011.

She contributed to the propulsion system of a series of early weather satellites known as Television Infrared Observation Satellite (TIROS); the rocket designs used in America's missions to the moon; the Atmosphere Explorer, the first upper-atmosphere rocket; and the Mars Observer that was lost in 1993 during the interplanetary cruise phase. She worked for NASA developing the rocket motor for the Space Shuttle from 1981 to 1983.

When Mrs. Brill died at age 88 in 2013, Michael Griffin, president of the American Institute of Aeronautics and Astronautics, praised her as "a pioneering spirit" who coupled "a clear vision of what the future of an entire area of systems should be with the ingenuity and genius necessary to make that vision a reality."

RACHEL FULLER BROWN & ELIZABETH LEE HAZEN

THE FIRST ANTIFUNGAL ANTIBIOTIC
INVENTION: NYSTATIN

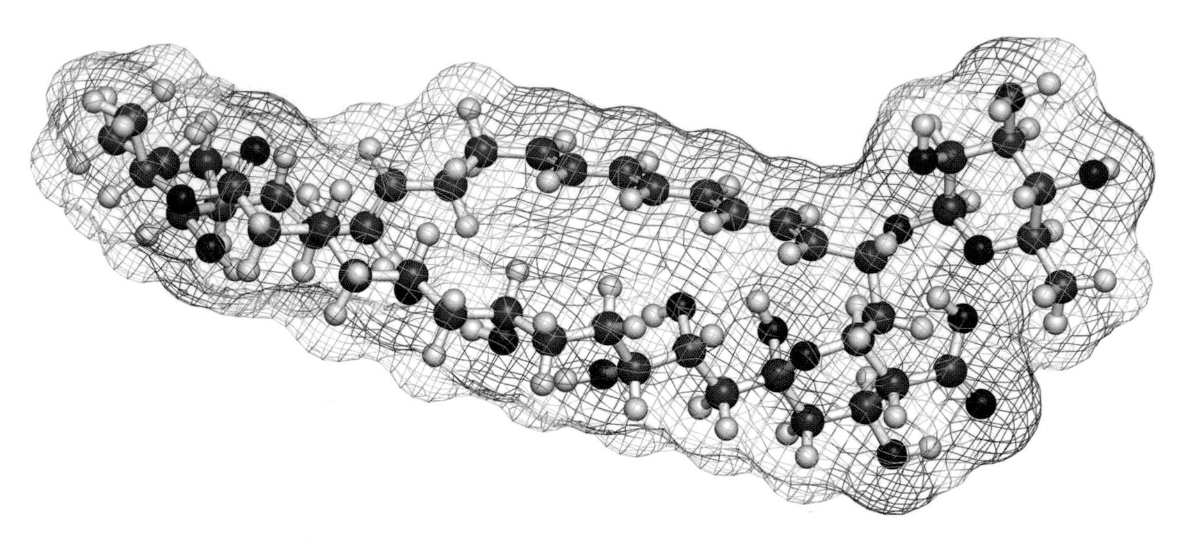

Nystatin antifungal drug molecule.

Rachel Fuller Brown (1898 – 1980) and Elizabeth Lee Hazen (1885 – 1973) owe their invention of Nystatin, the first antifungal antibiotic, to the efficiency of the US Postal Service whose sterling work enabled them to undertake a long-distance working relationship, Hazen situated in New York and her colleague working in Albany. Their relationship helped to end misery for countless numbers of people who suffered from fungal diseases and along the way it also contributed to the restoration of priceless art works and the treatment of Dutch elm disease.

Brown was born in Springfield, Massachusetts, in 1898 before moving to Missouri. Graduating from college in 1920, she went to the University of Chicago to take a master's in organic chemistry. She taught chemistry and physics for several years before going back to university to pursue a doctorate in organic chemistry.

Hazen was born in Rich, Mississippi, but was orphaned by the age of 3, living first with her grandmother and then her father's brother. She graduated from Columbia University with a master's in biology and completed a doctorate in microbiology at the university's College of Physicians and Surgeons. In 1931 she was running the Bacterial Diagnosis Laboratory at the New York Division of Laboratories and Research, where in 1944, she was selected to lead research into fungi.

TEAMING UP

In 1948, Brown teamed up with Hazen to work on the project that would make them both famous. Penicillin had been discovered in 1928 and antibiotics were now being used to treat a wide variety of diseases. There was an uncomfortable side-effect, however—the rapid growth of fungus that could cause painful or upset stomachs in patients. There were also fungal diseases that attacked the central nervous system, as well as conditions such as athlete's foot and ringworm for which there were, as yet, no cures.

The key to the discovery made by the two women lay, amazingly, in soil. Actinomycetes are microorganisms that live in soil and produce antibiotics. Hazen grew the organisms and tested whether they proved effective against two fungi—*Cryptococcus neoformans*, a fungus that caused the chronic disease cryptococcosis, a debilitating illness that affected the lungs, the skin, and the central nervous system; and *Candida albicans*, the cause of candidiasis which can be a fairly minor ailment as a vaginal yeast infection but which can also be serious in patients that have been treated with antibiotics.

ISOLATING THE AGENT

As soon as something in the soil looked like it had potential to cure these fungal diseases, Hazen mailed it to Brown, who isolated the active agent and returned the sample to Hazen to retest, before being tested on animals. Unfortunately, they all proved toxic to animals, until one particular sample, coincidentally from the vicinity of a farm owned by some friends of Hazen, proved not to be fatal to animals.

The name of her friends, Nourse, was given to this microorganism—*Streptomyces norsei*. It turned out to be effective not just against the two fungi they were researching, but also against fourteen others. The drug that resulted from this work was called Nystatin, in tribute to the New York State Division of Laboratories and Research under whose aegis Brown and Hazen made their discovery.

Brown and Hazen waived their rights to the royalties from Nystatin—amounting to $13.4 million—and instead gave them to the Philanthropic Research Corporation and to their own Brown-Hazen Fund to support scientific research and encourage women to take up careers in science.

EARLY PATENTS IN THE UNITED STATES

In 1809, the first patent ever granted to a woman—Mary Dixon Kies—was issued by the United States Patent Office in Washington. It was at the time totally unprecedented and was covered widely in the press. Women had certainly invented things in the past but had never taken it so far as to apply for a patent, or, occasionally, the patent was taken in the name of their husbands. From 1809 to 1829, eleven patents were issued to women and in the following 20 years another twenty-two. It is worth noting that the patents applied for were not purely for inventions that related to women but for a range of articles including foot stoves, a sawing device, bellows for a stove, a machine for cutting straw, a process for extracting fur from skins and turning it into yarn, a fireplace, improvement in the method for applying distemper paint, a submarine telescope and lamp, and a method for making sad irons. By 1895, 5,200 patents had been issued to women and in the ensuing fifteen years, no fewer than 3,615 patents were issued to women.

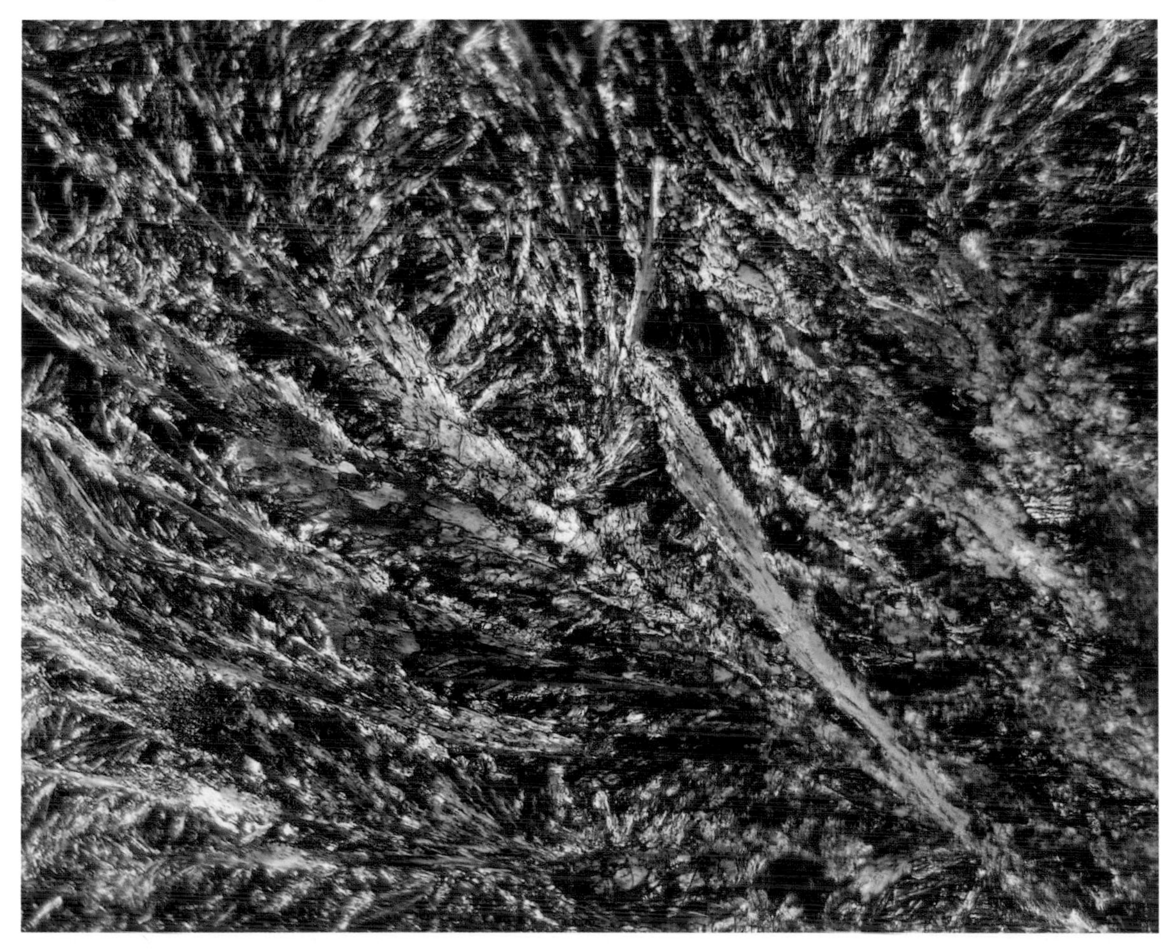

Nystatin antifungal drug as seen under a microscope.

MAYA BURHANPURKAR

TEEN SCIENCE WHIZKID
INVENTION: AN INTELLIGENT ANTIBIOTIC

Maya Burhanpurkar spent a year developing software for a low-cost, self-driving technology for power wheelchairs.

Canadian researcher Maya Burhanpurkar was born in 1999 in Orillia, Ontario, and in 2017 enrolled at Harvard University. This precocious young woman is already a scientific star, however, having at the age of 12 developed a prototype for an intelligent antibiotic. An intelligent antibiotic is one that selectively kills pathogenic bacteria, such as *E. coli* but while doing so it leaves other bacteria intact.

At the age of 10 this scientific prodigy built a microbiology laboratory in the basement of her family home. She had already volunteered at a hospital in India and began conducting her own research. At the age of 13, she received the Platinum Award at the Canada-Wide Science Fair for the work she had been doing on the cardiac and gastrointestinal safety of Alzheimer's drugs. She had a personal interest in this as her grandfather had died of Alzheimer's.

Maya continued to receive awards and in 2013, she was named one of Canada's Top Twenty Under Twenty, the highest civilian honor for a young Canadian person. This extraordinary young woman has also been awarded the Queen Elizabeth II Diamond Jubilee Medal and all the signs are she has a great future ahead of her.

HERMINIE CADOLLE

REMODELING WOMEN'S UNDERWEAR

INVENTION: THE MODERN BRA

At the end of the nineteenth century, Paris was the center of the fashion world, but Herminie Cadolle (1845 – 1926), a retailer and designer of lingerie, decided to leave the French capital behind and travel to Argentina to seek her fortune. She may also have decided to get out of town because she enjoyed a close friendship with the notorious French anarchist Louise Michel, described by one observer as the "French grande dame of anarchy."

At that time, Argentina was a rapidly expanding new nation of South America and Herminie opened a lingerie boutique in its capital, Buenos Aires. Before long, her boutique was the hub of the city's social life, a meeting place for the fashionable wealthy elite. She traveled often to France, bringing back French seamstresses who could teach their skills to Argentinean women. Business boomed and she extended her boutique and opened others.

The Exposition Universelle of 1900 was held in Paris to demonstrate the very latest in technology, fashion, and manufacturing. It was here that Herminie first demonstrated her new invention. It was a simple idea but it would revolutionize women's undergarments.

In order to greatly increase comfort, she cut the traditional corset in two, in effect creating the first bra. She patented it, describing it as the "corselet gorge" and marketing it with the brand name "Le Bien-Etre"—the "Wellbeing." The lower part of the garment was a corset for the waist and the upper part supported the breasts, as in a modern bra, with the use of shoulder straps. By 1905, the upper part was being sold independently of the corset and was known as a "soutien-gorge" which literally means "support for the throat" although "gorge" was also an archaic French word for breast. Bras are still known in France as "soutiens-gorges."

Soon, the world was flooding to Herminie's shops and she supplied bras to the great names of the day—queens, dancers, actresses, and even the exotic dancer Mata Hari was one of her customers. Herminie Cadolle died in 1926, at age 81, but her business continued and exists to this day.

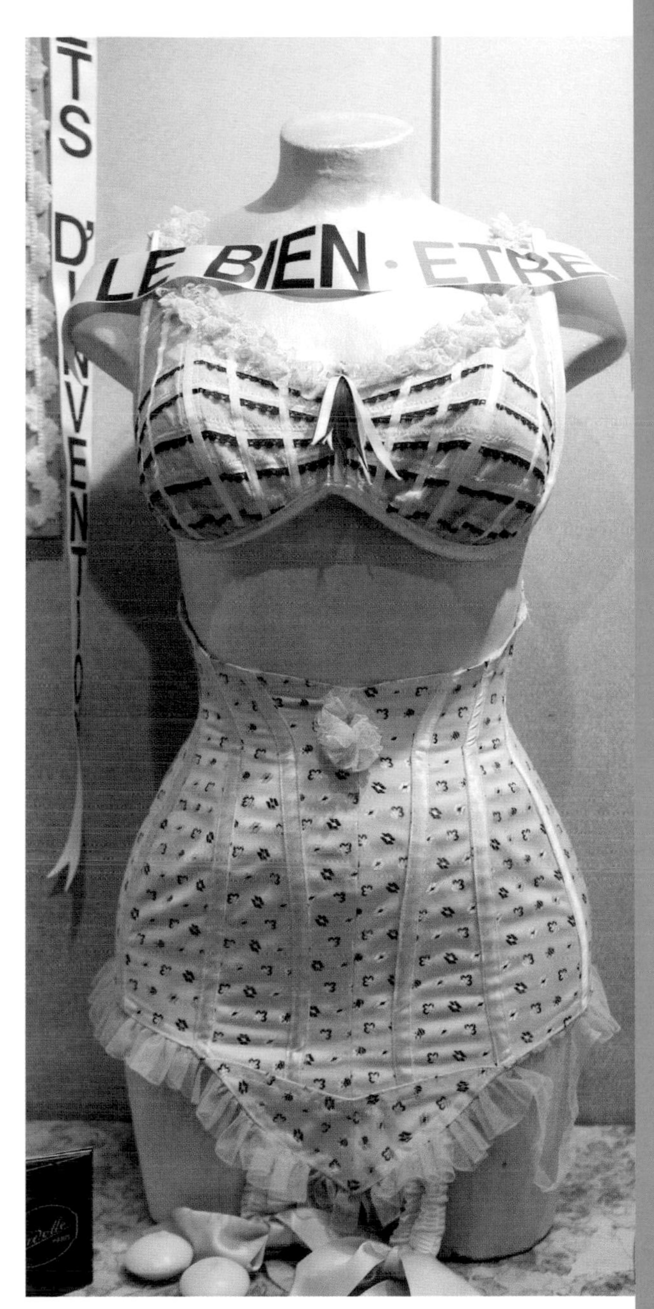

Herminie Cadolle invented the "soutien-gorge" brassiere in 1889.

MARY P. CARPENTER

A STITCH IN TIME

INVENTION: SEWING MACHINE MODIFICATIONS

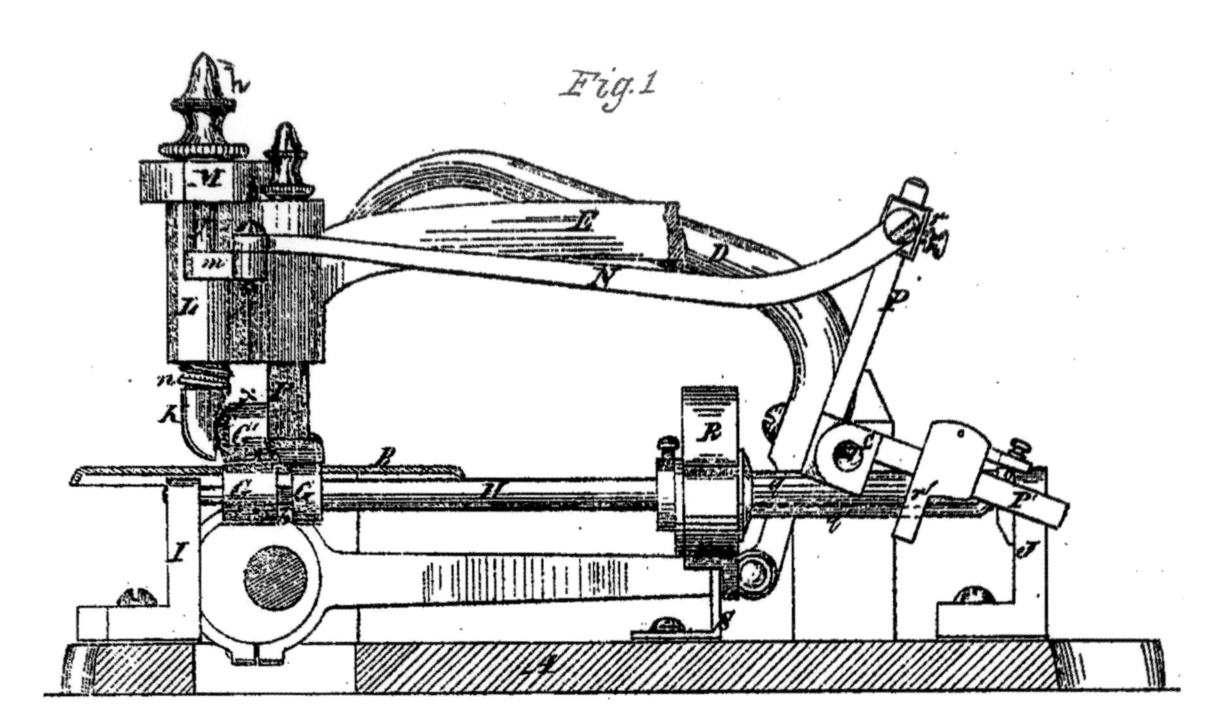

Patent 112,016 by Mary P. Carpenter, February 21, 1871, for improvements to sewing machine feed mechanism.

Probably born in the 1840s, Mary Carpenter should be lauded by homemakers everywhere as a woman dedicated to saving time in everyday domestic chores. Indeed, she held thirteen patents that were devised to do just that.

She was brought up in Buffalo, NY, and seems to have been taught mechanical drawing. She also knew a great deal, it seems, about how textiles were manufactured. No record of her birth has ever been found in Buffalo and it is speculated that she was, in fact, a Scottish immigrant.

She was awarded her first patent in 1862, for an ironing machine that was designed to simplify the method of creating fluted pleats in a garment. This was a style that was particularly fashionable at the time and a machine to make it easier to achieve that look must have been very welcome.

In 1871, she devised an improvement in the feed-motion of the sewing machine and a year later she was granted another sewing-machine patent, this time for "sewing-machines which are especially designed for sewing straw-braid and consists mainly in the construction of the hooked needle employed for drawing the loop through the braid, and also in the mechanism for giving certain peculiar movements to said needle."

1885 brought her a patent for a type of coal shovel and one for a mosquito-net bed canopy arrived that same year. She must have been particularly bothered by mosquitoes where she lived because in 1890 she was granted a patent for a mosquito trap. In the next few years, she earned a patent for an embroidery holder and a device to improve the hang of certain skirts with folds. A patent for a hair comb was her last, granted in 1904.

JOYCE CHEN

STIR-FRY STAR CHEF

INVENTION: THE FLAT-BOTTOM WOK WITH HANDLE

Joyce Chen's trailblazing 1962 cookbook introduced US diners to Chinese ingredients for the first time.

Joyce Chen was born in Beijing in 1917, but by 1949 China had been embroiled in civil war for many years, so Joyce, her husband Thomas, and their two children fled to the United States, settling in Cambridge, Massachusetts. Her mother and her governess were in charge of the kitchen, but Joyce had always been interested in cooking, and she watched them closely, to learn the secrets of Chinese cuisine.

In 1958, she opened a restaurant in Cambridge, naming it simply the "Joyce Chen Restaurant." She adopted an innovative approach, creating the all-you-can-eat Chinese buffet to give sales a boost on Tuesday and Wednesday nights when business was slow. Customers could try unfamiliar but authentically Chinese dishes, such as Peking duck, moo shu pork, hot and sour soup, and potstickers—a kind of Chinese dumpling that she dubbed "Peking ravioli" or just "ravs."

At her second restaurant, "The Joyce Chen Small Eating Place," which opened in 1967, Joyce introduced the concept of *dim sum*—the Chinese style of eating bite-size portions served in small steamer baskets or on small plates. She opened two further restaurants.

At the same time, she was also teaching Chinese cooking and she published a cook book, the *Joyce Chen Cook Book* that became extremely influential. Television appearances followed and she had a show that was broadcast in America, Britain, and Australia.

Her invention in 1971 of the flat-bottom wok—also known as a stir fry pan—with a handle helped popularize Chinese cooking and featured as one of a range of cooking utensils that she launched. Joyce Chen died in 1994 but in 2014 she was one of five chefs featured on the "Celebrity Chefs Forever" series of stamps issued by the US Postal Service.

EDITH CLARKE

ANALYZING ELECTRIC POWER
INVENTION: THE CLARKE CALCULATOR

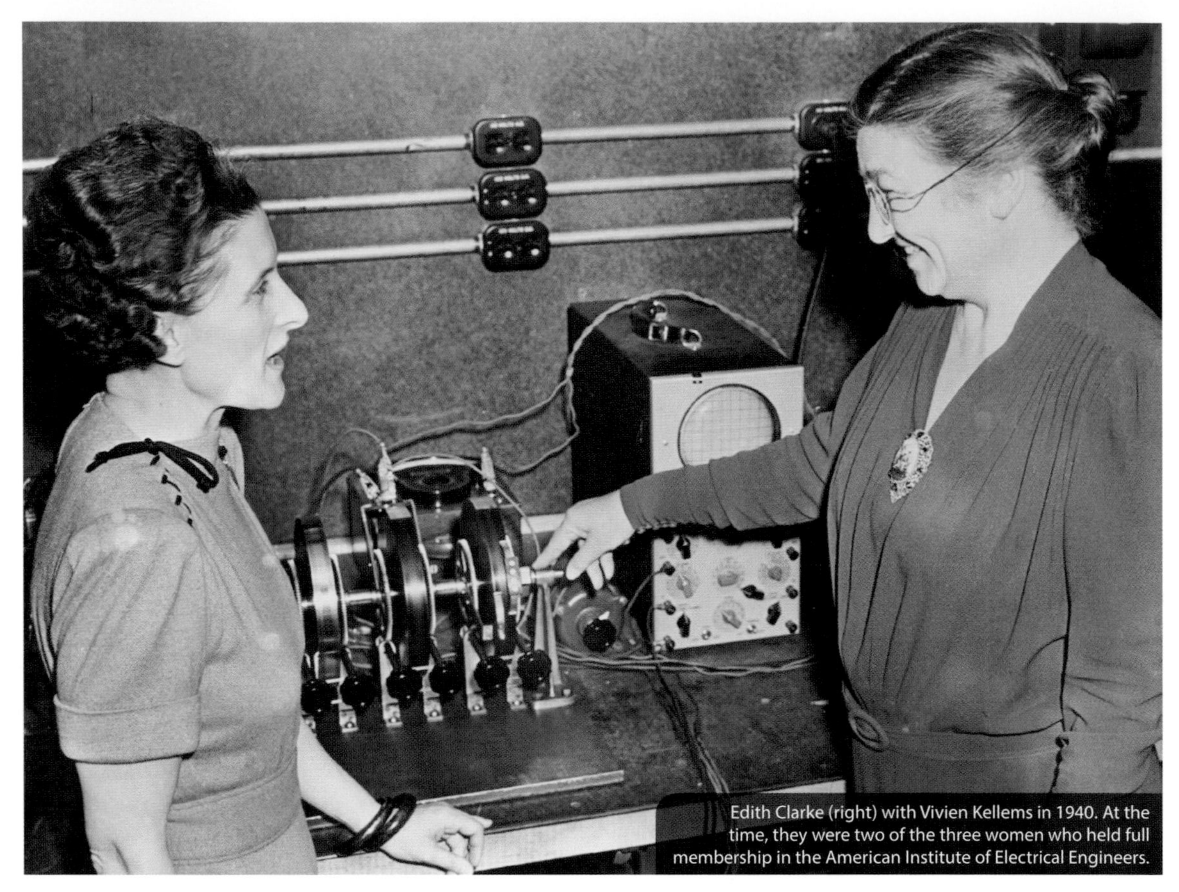

Edith Clarke (right) with Vivien Kellems in 1940. At the time, they were two of the three women who held full membership in the American Institute of Electrical Engineers.

Born in a small farming community in Maryland, Edith Clarke (1883 – 1959) studied mathematics and astronomy at Vassar College, graduating with honors in 1908. For a couple of years she taught mathematics but in 1911 she enrolled at the University of Wisconsin as a civil engineering student. During her first year at university, however, she took a summer job at AT&T as a "Computor Assistant" to research engineer Dr. George Campbell. She enjoyed the work so much that she stayed on at AT&T to train and manage a group of (human) computors.

She returned to her studies in 1918, enrolling in the electronic engineering course at the Massachusetts Institute of Technology and was the first woman to graduate with an MSc degree. Unable to find work as an engineer—women were not taken seriously in the profession—she found employment supervising a team of "computors" for General Electric in Schenectady, New York. She worked in the Turbine Engineering Department specializing in electrical power system analysis.

Edith Clarke is remembered for the Clarke Calculator invented in 1921 and patented in 1925. It was a simple graphic device that was used in solving electric power transmission line problems. It simplified the equations electrical engineers employed to understand power lines. It could solve equations involving hyperbolic functions ten times faster than had been achieved before.

In February 1926, Clarke was the first woman to present a paper at the American Institute of Electrical Engineers. In 1947 she became Professor of Electrical Engineering at the University of Texas at Austin, the first female professor of electrical engineering in American history. She retired in 1957 and died two years later, at age 76.

ADELAIDE CLAXTON

AVOIDING OUTSTANDING EARS

INVENTION: AN EAR-CAP

The Claxton ear-cap patented by Adelaide Claxton to correct "outstanding ears" and worn by children when asleep.

> *It has often been observed by experienced elders, that since it became the fashion for babies to discard caps, protruding ears are but too common. They are very ugly, and the ear-cap just invented is a safe preventive, without the heat that made the cap objectionable.*

In 1891, this quote from the *Northampton Mercury* was trumpeting the invention of an ear-cap for babies by the artist Adelaide Claxton (1830 – 1879). It was fastened with a delicate web of ribbons that was designed specifically to prevent babies' ears from getting pushed forward as they slept and sticking out when they woke up. The ear-cap would keep them tight against the side of the head, no matter what position they slept in. Claxton must have been hopeful of success because she patented it in both Britain and the United States.

Claxton was a late nineteenth-century British painter and illustrator who made a living mostly through providing illustrations to the commercial press and selling satirical illustrations to more than a dozen magazines and periodicals such as *The Illustrated London News*, *London Society*, and *Judy*. Born in London, she was the daughter of another British painter, Marshall Claxton, and painted, mostly in watercolors, domestic scenes with literary or fantasy elements.

MRS. CLEMENTS

TICKLING THE TASTEBUDS

INVENTION: ENGLISH MUSTARD

Mustard powder is obtained by grinding mustard seeds. To make a powder, toast your mustard seeds for 20 seconds in a dry skillet. Cool the seeds, then transfer to a spice grinder and pulse until you have a powder.

Prior to 1729, mustard was made from mustard seeds that were pounded into a powder and then boiled in vinegar. It is said that an elderly lady named Mrs. Clements who lived in Durham in the north of England, invented dry English mustard, grinding the seeds to a fine powder in a mill and selling the resulting product. She is reported to have made a small fortune from her invention, selling it off the back of a packhorse and even securing the patronage of King George I. Strictly speaking, as it was not mixed with vinegar, it was no longer really mustard, but it retained that name and has remained popular ever since.

ELEANOR COADE

A REMARKABLE NEW BUILDING MATERIAL INVENTION: COADE STONE

The Coade stone lion at the southern end of Westminster Bridge, London, is very close to the site of the original Coade's Gallery showroom.

Lithodipyra (literally "stone fired twice") was a very hard-wearing, weather-resistant ceramic stone used to mold Neoclassical statues and decorative stoneware that exists to this day. It is of very high quality, unlike many of its precursors that had been patented during the previous decades. Eleanor Coade (1733 – 1821), a British businesswoman, created her own unique version of this stone, known as Coade stone. She had perfected the clay recipe and the firing process to such an extent that her improved stone was streets ahead of any competitors.

Born in Exeter, Devon, she was the daughter of a wool merchant from Lyme Regis. In 1769, 36-year-old Eleanor purchased an artificial stone factory in Lambeth, London. Coade's Artificial Stone Manufactory made and sold Coade's Lithodipyra for the next five decades.

She worked closely with a talented sculptor, John Bacon, and made him works supervisor. Soon, they were winning contracts from all the major architects of the Georgian age, such as John Nash and Robert Adam, and she hired eminent designers and modellers to work her stone.

Mrs. Coade, as she was generally known, was a consummate businesswoman. In 1784, she created a catalog that featured no fewer than 746 designs featuring busts, statues, coats of arms, and a wide range of architectural ornaments. She opened Coade's Gallery, a showroom at the southern end of Westminster Bridge. Her stoneware still graces some of Britain's most iconic buildings, such as St. George's Chapel at Windsor, the Royal Pavilion in Brighton, Carlton House in London, the Royal Naval College at Greenwich, and Buckingham Palace.

Following her death in 1821, the business eventually collapsed in 1844, partly because the fashion for such stone was ending but also because of a large debt owed to the firm by the Duke of York.

JOSEPHINE COCHRANE

ALL WASHED UP

INVENTION: THE AUTOMATIC DISHWASHER

The inventor of the first commercially successful dishwasher was, as one might have predicted, a woman. Born in Ashtabula County, Ohio, Josephine Cochrane (1839 – 1913) was the daughter of a civil engineer. Josephine married William Cochran (she added an "e" to her surname after her husband's death) in 1858, who became a wealthy dry goods merchant and Democratic politician.

Their prosperity gave the Cochranes a mansion, servants, and a busy social life, and they threw regular dinner parties. After one of these events, Josephine was very disappointed to discover that the servants had chipped some of her expensive crockery while washing up. She resolved to find a safer alternative and also to remove some of the drudgery from housewives' daily lives.

Her life was irrevocably changed in 1883 when her alcoholic husband died, leaving her in debt with just $1,535.59 in the bank. It forced her to persevere with her development of a dishwasher. With the help of a mechanic named George Butters, she created wire compartments to fit the size of her dishes and there were specific compartments for plates, side plates, cups, and saucers. These were placed inside a wheel that lay flat inside a copper boiler. The wheel was turned by a small motor and hot, soapy water was squirted up from the bottom of the boiler before pouring down on the dirty crockery. It was a novel approach. Other machines had used scrubbers inside the boiler and unpressurized water.

She patented her dishwasher in 1886 and demonstrated it at the 1893 World's Columbian Exposition in Chicago where she was awarded the highest prize for "best mechanical construction, durability and adaptation to its line of work." Soon, her machine was being ordered by restaurants and hotels all over Illinois and she went into production with the opening of a factory, Garis-Cochran Manufacturing Company.

Josephine Cochrane continued selling her dishwashers almost until her death in 1913 at the age of 73. Her company was bought by Hobart which became KitchenAid and exists today as Whirlpool Corporation. Cochrane was posthumously inducted into the National Inventors Hall of Fame in 2006 for her invention.

Garis-Cochrane Dish Washing Machine (1909) sales advertisement.

THE WORLD'S COLUMBIAN EXPOSITION

The World's Columbian Exposition, also known as the World's Fair, was held in Chicago in 1893 in celebration of the 400th anniversary of Christopher Columbus's arrival on the shores of the New World. The fulcrum around which the fair spun was a large lake of water that was there to represent the long sea voyage Columbus had made. The fair, won in competition with New York City, Washington DC, and St. Louis, became a hugely influential social and cultural event, its impact being felt on architecture, the arts, electricity, and sanitation. It also provided a huge boost to American industrial optimism.

Designed by John Wellborn Root, Daniel Burnham, Frederick Law Olmsted, and Charles B. Atwood, its layout was aimed to be a prototype of what its creators thought a city should be, employing the Beaux Arts principles of French neoclassical architecture. Its fourteen main buildings were designed by eminent architects and the exhibits featured well-known artists and musicians.

Covering more than 600 acres, it had 200 temporary buildings as well as canals and lagoons. Forty-six countries were represented in its stands and buildings and during the six months it ran, a staggering 27 million people visited. Attractions included a Ferris wheel, carnival rides, life-size replicas of Columbus's three ships, the first moving walkway or travelator, and a beautiful chapel designed by Louis Tiffany.

WOMEN INVENTORS AT THE WORLD'S FAIR

Many women inventors grabbed the opportunity to showcase their inventions at the fair, among the inspired innovations were: a collapsible noiseless coal-scuttle; a combined dress stand and fire escape; drawings of inventions for extracting gold from base metals; a folding mail cart; a combined traveling trunk and wardrobe; protectors for fingers when sewing; a hygienic egg boiler; an expansible umbrella holder; and an improved ear trumpet.

This illustration appeared as the frontispiece of the catalog for Art and Handicrafts in the Women's Building at the World's Columbian Exposition, 1893.

ANNA CONNELLY

REACHING FOR THE SKY
INVENTION: THE FIRE ESCAPE BRIDGE

Colorful Soho building facades with painted fire escapes, Manhattan, New York City.

By the nineteenth century in America's cities, apartment buildings were adding floors, factories were being built higher with multiple stories and public buildings, too, were reaching for the sky. Often, these buildings were made of wood, turning them into death-traps in the event of fire. Firefighters did their best but at the time their ladders, at most, reached only to the fourth floor of a building. The lives of people on higher floors were in grave danger.

There was a public outcry and inventors began to come up with ways for people to escape from fires. One man even patented a head-mounted parachute that was designed to let the wearer float to the ground. Finally, the legislature responded by passing a law in 1861 requiring fire escapes to be part of all newly constructed tall buildings and houses.

In 1871 a further law was passed requiring the retroactive fitting of fire escapes to older buildings, hotels, offices, and factories. But the fire escapes of the nineteenth century were far from ideal. In some cases, they were unable to cope with the number of people trying to get down them.

Anna Connelly came up with a clever solution. She is often wrongly credited with the invention of the fire escape, but what she actually invented was a fire escape bridge. In the event of a stairway blocked by flames and smoke sometimes the only safe direction of travel was up. Thus, in 1887, Connelly, about whom little else is known, was granted a patent for an iron-railed bridge. Instead of trying to escape down the stairs, people had to find their way up to the roof of the burning building and cross to the safety of the next building using the bridge.

Connelly's invention had the added benefit of allowing firefighters to more effectively target fires by hauling water to specific areas of a building. The bridge was also extremely cost effective as it was merely added to the exterior of the building, eliminating the need for costly interior redesign. Throughout the years following it was responsible for saving many thousands of lives.

MARTHA COSTON

PYROTECHNIC LIFE-SAVERS
INVENTION: THE COSTON FLARE

Coston's Telegraphic Night Signals as used by the US Army Signal Service in 1864.

Martha Coston (1826 – 1904) was born Martha Hunt in Baltimore, Maryland, but was living in Philadelphia when she eloped at the age of 15 with Benjamin Coston, a 21-year-old inventor. Coston was appointed director of the US Navy's scientific laboratory in Washington DC, where he developed a number of devices including a signaling rocket. At the time ships communicated using flags during the day and lanterns at night. In 1847, Coston was working on a system of color-coded night signals when he left the Navy over a pay dispute, taking a job with the Boston Gas Company. Sadly, his health suffered as a result of working with chemical fumes and he died the following year. Martha was left with four children at age 21, and no income.

While putting her late husband's papers in order, she discovered his notes on night signaling. For the next ten years she worked on the idea. It was a struggle until, in 1858, she went to a fireworks display in New York City. Watching the fireworks explode in the night sky, it suddenly struck her that she needed a blue flare to go with the red and white ones that she had already developed. It was the final piece of the jigsaw puzzle. Ultimately she had to settle on a green flare, but it still worked the way she wanted. In April 1859, she was granted a patent for a pyrotechnic night signal and code system—or rather her deceased husband was granted a patent as the inventor and she was named as "adminastratrix."

The system was basically a type of semaphore, the colored flares being burned in various combinations to represent the numerals zero to nine and the letters A and P. Coston's company produced more than 100,000 sets for the US Navy during the American Civil War alone and it was adopted by many other navies, including Brazil, Denmark, France, Italy, and the Netherlands. In 1871 Martha was granted patents for improvements to her system—this time in her name alone.

Her sons inherited the business when Martha died in 1904, at age 78, and remained in business until at least 1985. Martha Coston was probably one of the most successful of the nineteenth-century's female inventors and she was inducted into the National Inventors Hall of Fame in 2006.

CARESSE CROSBY

CHANGING THE SHAPE OF WOMEN
INVENTION: THE MODERN BRASSIERE

Caresse Crosby's Backless Brassiere.

Caresse Crosby had numerous affairs, three husbands, was a champion of art and literature, and was friend and supporter of the writers and artists of the Lost Generation in Paris. Almost incidentally, she was also the inventor of the modern bra, changing the shape of women around the world forever.

Her life was the stuff of novels and films. Born Mary Phelps Jacob in New Rochelle, New York, in 1891, she was from a wealthy New England background, her parents were both descended from American colonial families. It was fairly inevitable, therefore, that she would marry someone from a similar background. Richard R. Peabody was a blue-blooded Bostonian whose family had crossed the Atlantic and landed in New Hampshire in 1635.

A TORRID ROMANCE AND AN OPEN MARRIAGE

Mary and Richard Peabody had two children, but Richard took to the bottle when he came back from fighting in World War I. At 29, Mary's marriage was falling apart so she began a romance with another young man, the handsome, passionate Harry Crosby. Their torrid relationship caused much tongue-wagging among Boston's elite, but for her it was wonderful.

In 1922, after she divorced Peabody and married Crosby, the pair moved to France where, at his behest, she changed her name to Caresse. The couple fitted into the bohemian life of Paris perfectly, indulging in the alcohol, the opium, the wild parties, and the sex. Theirs was an open marriage long ahead of the trend for such things. Among the men Caresse bedded was the great photographer, Henri Cartier-Bresson, with whom she had an intense relationship.

In 1927, Caresse and Crosby launched their own publishing house, through which they published their own poetry, and the work of others including D.H. Lawrence, James Joyce, and Henry Miller. Sadly, however, Harry was soon dead. He shot himself in the head as part of a suicide pact with his lover Josephine Noyes Rotch who was found dead beside him.

INVENTING THE BRA

How did Caresse Crosby find time to invent the modern bra in among the sex, drugs, and alcohol? It happened when she was 19 years old, and still known as Mary Phelps Jacob. In 1910 the standard undergarment for a woman was the whalebone corset, a lumpy and very uncomfortable item of clothing that destroyed the line of a dress, especially if it was sheer like the one she was wearing on this particular night as she prepared to attend a debutante's ball.

When she looked in the mirror she did not like what she saw and shouted to her maid to bring her a couple of silk handkerchiefs, a cord, some ribbon, and a needle and thread. Sitting down, she began to sew, creating as she did so the basis of the first modern bra.

Later, when she made her entrance to the ball she was immediately surrounded by friends wanting to know what on earth she was wearing, and where could they buy one? They even offered her money for one. It suddenly struck her that this item that she had named the "brassiere" actually had potential. She applied for a patent for her "Backless Brassiere" and US Patent 1,115,674 was granted to M.P. Jacob on November 3, 1914, the first bra patent in the United States. Her application described her invention as "capable of a universal fit to such an extent that ... the size and shape of a single garment will be suitable for a considerable variety of different customers." She added that it was "so efficient that it may be worn even by persons engaging in violent exercise like tennis."

THE FASHION FORM BRASSIERE COMPANY

It was revolutionary, a much more comfortable garment than the corset which pinched women's waists and crushed their ribs but Caresse's bra did not provide much in the way of support. It separated the breasts but also flattened them. The timing was perfect as when World War I started, the United States War Industries Board banned corset manufacture because the metal for the ribbing was required for the construction of battleships.

In 1922 she established the Fashion Form Brassiere Company in Boston, and took on a couple of women to manufacture her bras. It did not last long, however. She produced no more than a few hundred and secured some orders from shops but it never really took off and anyway, her lover Harry Crosby had a disdain for commerce.

Harry also had a generous trust fund which removed any real desire to make a go of it. She closed the business and later sold the rights to her invention to the Warner Brothers Corset Company of Bridgeport, Connecticut, for $1,500. They went on to earn $15 million from her patent during the ensuing thirty years.

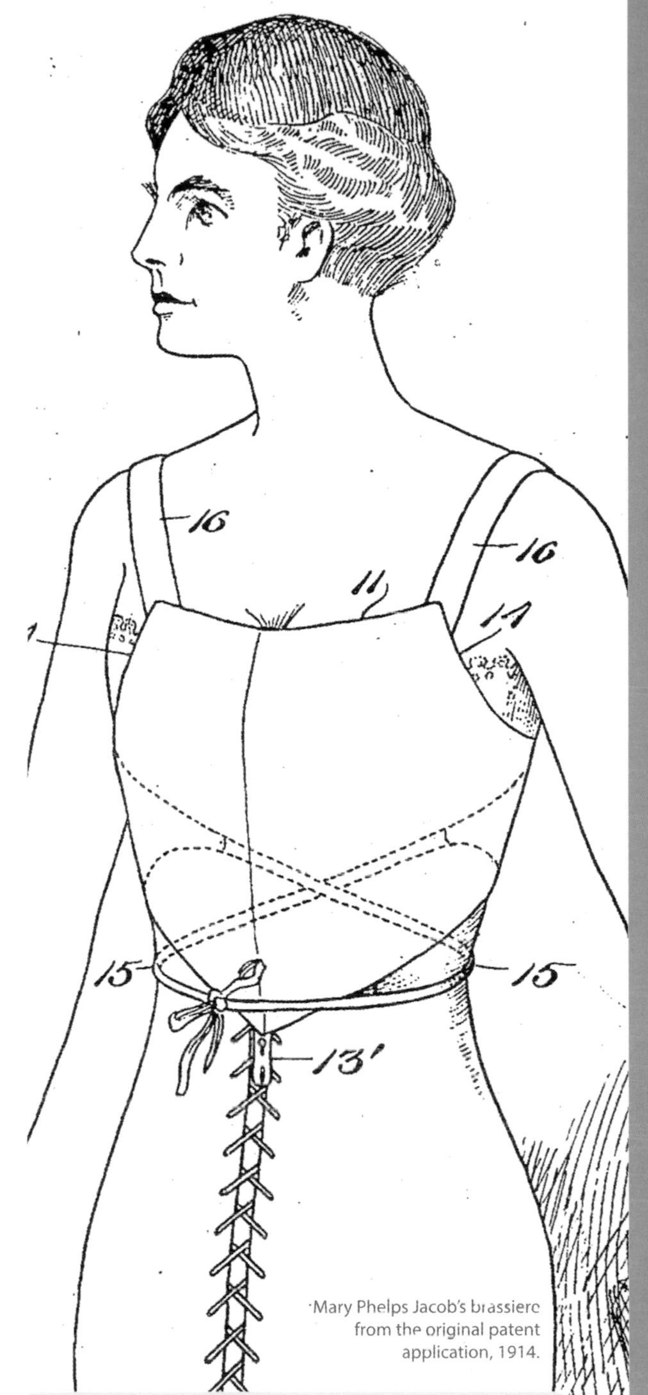

Mary Phelps Jacob's brassiere from the original patent application, 1914.

DO MEN KNOW "BREAST?"

When it comes to the invention of the bra, there are several men who definitely wanted a slice of the pie. It just makes sense that men would naturally be interested in anything related to bosoms. Henry S. Lesher of Brooklyn, New York, patented a bra-like device in 1859. But almost certainly Henry never tried it on. It was a complicated, confusing apparatus, and very uncomfortable looking— and it came with built-in "Arm Pit Shields" to protect against perspiration!

EMILY CUMMINS

A FRIDGE FOR AFRICA
INVENTION: A SUSTAINABLE REFRIGERATOR

Emily's gap-year trip to Namibia meant she could improve her prototype solar-powered fridge which is increasingly being used in developing countries.

English inventor Emily Cummins, who was born in 1987, reasoned that electrical appliances powered by fossil fuels had to be a thing of the past. So, she invented a sustainable fridge ingeniously "powered" by dirty water. She describes her prototype as consisting of two metallic cylinders, one inside the other with a locally sourced material packed tightly into the space between them. This could be sand or wool, for example, and it is soaked with water after it has been packed in.

When the fridge is situated in a warm place, the energy of the sun makes the outer part of the fridge "sweat." The water in the material evaporates and the heat is transferred away from the inner cylinder, inside which is the food to be kept fresh, and this inner cylinder becomes cooler.

Cummins' fridge is ideally suited for use in the developing world as it requires no electricity and can be built using everyday items such as barrels or car parts. She has given the design away in townships across southern Africa.

In 2007, Cummins was named British Innovator of the Year, and in 2010 she was honored by Junior Chamber International as one of the Ten Outstanding Young Persons of the World.

EMILY'S SUSTAINABLE FRIDGE

Emily Cummins has developed a way of using the sun's power to help impoverished communities in Africa. Her eco-friendly, sustainable fridge is based on a simple principle: it uses the sun's rays to evaporate water, which in turn keeps the contents cool.

HOW DOES IT WORK?

1. The fridge is made up of two cylinders, one inside the other. It is not connected to any power source. The outer cylinder is made of any solid material (e.g. wood or plastic) with holes drilled in the side.
2. The inner cylinder is made of metal and has no holes to ensure its contents remain dry.
3. The gap between the inner and outer cylinders is filled with material such as sand, wool, or soil, that can be soaked with water.
4. In hot weather the sun's rays heat this wet material and water evaporates off. As the material is held against the inner cylinder wall, heat is removed from the inner chamber by the evaporation process, keeping it at a cool temperature of 6°C.
5. Re-soaking the material with more water will keep the fridge working.

Emily's electricity-free fridge is now used across South Africa, Zambia, Zimbabwe, and Botswana, making a real difference to people's lives. Unlike previous pot-in-pot coolers, the contents are kept dry and hygienic because the water does not come into contact with the product.

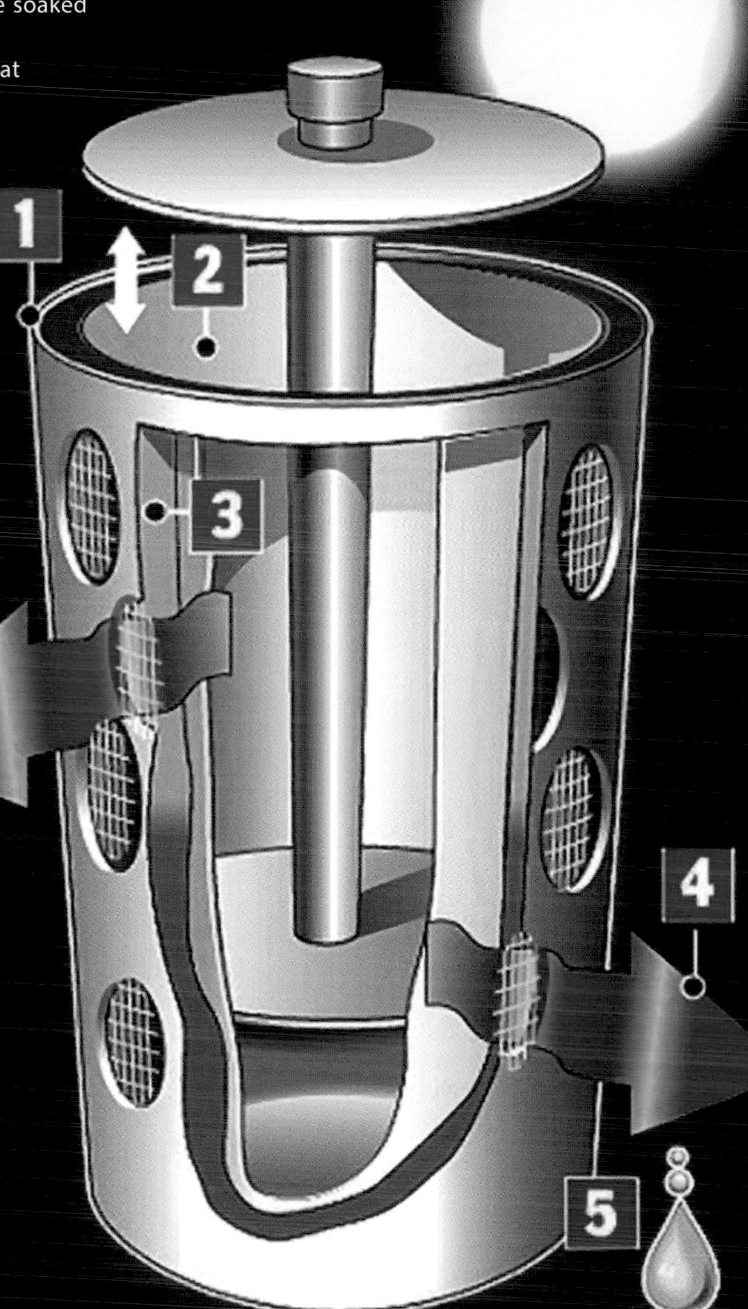

MARIE CURIE

THE QUEEN OF SCIENCE
INVENTION: POLONIUM, RADIUM, AND X-RAY TECHNOLOGY

Marie Curie giving a lecture on radioactivity at the Academy of Medicine, 1925.

Born Maria Skłodowska in 1867 in Warsaw, Poland, she was the daughter of teachers who had four other children. The family had been dispossessed of its property and money due to its involvement in Polish national uprisings against the Russians which condemned the children to a difficult life, as they just made ends meet. Eventually, Maria's father, a teacher of maths and physics, was fired and forced to take positions that did not pay well and her mother died when she was only 10 years old. Maria and her sister were not permitted to enroll for higher education because of their gender but they became involved with the Flying University, a Polish institution that admitted women students.

Maria made an arrangement with her sister, Bronislawa. She would help fund her medical studies in Paris by working as a governess for two years if Bronislawa would do the same for her. She worked for about two years until Bronislawa invited her to join her and her new husband in Paris but rejected the invitation because she still did not have sufficient funds. Finally, in 1891, she traveled to Paris, enrolling at the Sorbonne where she obtained degrees in physics and the mathematical sciences. Meanwhile, she had met and fallen in love with Pierre Curie, a professor of physics. The two were married in 1895. A contemporary quipped that Marie was "Pierre's biggest discovery."

Pierre and Marie Curie in their laboratory.

THE NEW PHENOMENON

Life was tough and the Curies' pioneering early research into the invisible rays given off by uranium was conducted in dangerous conditions in a shed next door to the School of Physics and Chemistry in Paris. Evidence of rays that could pass through solid matter had been discovered in 1896 by the French physicist Henri Becquerel. The Curies were inspired by his work and called the new phenomenon radioactivity.

Eventually, their research led to the discovery of a new chemical element they named polonium (after Poland). After they had extracted polonium, the residual liquid was still extremely radioactive, and they suspected it must contain another substance. In 1898, the Curies discovered that hidden substance, a new element which they called radium. Throughout her life Marie Curie was an advocate of radium's healing properties, despite the radiation burns she experienced during experiments. In 1903, Pierre and Marie Curie were awarded the Nobel Prize for Physics with Henri Becquerel for their joint research on radioactivity.

MODEST AND DIGNIFIED

In 1906, Pierre Curie died in a street accident in Paris. Modest and dignified, Marie Curie succeeded him as Professor of General Physics in the Faculty of Sciences at the Sorbonne in Paris, the first woman to hold that position. She was awarded a second Nobel Prize, this time for Chemistry, in 1911, in recognition of her work in radioactivity, and many prestigious prizes followed throughout her career.

During World War I she created mobile x-ray units known as the "Little Curies." She put x-ray machines into cars that were driven onto the battlefield where the x-rays could be carried out on the injured. This kind of battlefield treatment was many years ahead of its time.

Marie Curie died of leukemia in Savoy, France, in 1934, at age 65. The Curies both experienced severe radiation sickness, and even now, over 100 years later, researchers examining their papers have to wear protective clothing.

WHAT IS FLUOROSCOPY?

Fluoroscopy is an imaging technique that uses x-rays to examine moving images in the body, such as pumping action of the heart or the motion of swallowing. It is similar to radiography and x-ray computed tomography (x-ray CT) in that it generates images using x-rays, but the difference is that radiography shows fixed still images on film whereas fluoroscopy provides live moving pictures.

Fluoroscopy's origins and radiography's origins can both be traced back to the scientist Wilhelm Röntgen who noticed fluorescence during his discovery of x-rays in 1895. In the late 1890s, Thomas Edison began investigating materials for ability to fluoresce when x-rayed, and by the turn of the century he had invented a fluoroscope with sufficient image intensity to be commercialized. No precautions against radiation exposure were taken in those days as its hazards were not known at the time. Edison himself damaged an eye in testing these early fluoroscopes.

Fluoroscopy is also used in contemporary airport security scanners to check for hidden weapons or bombs. These machines use lower doses of radiation than medical fluoroscopy.

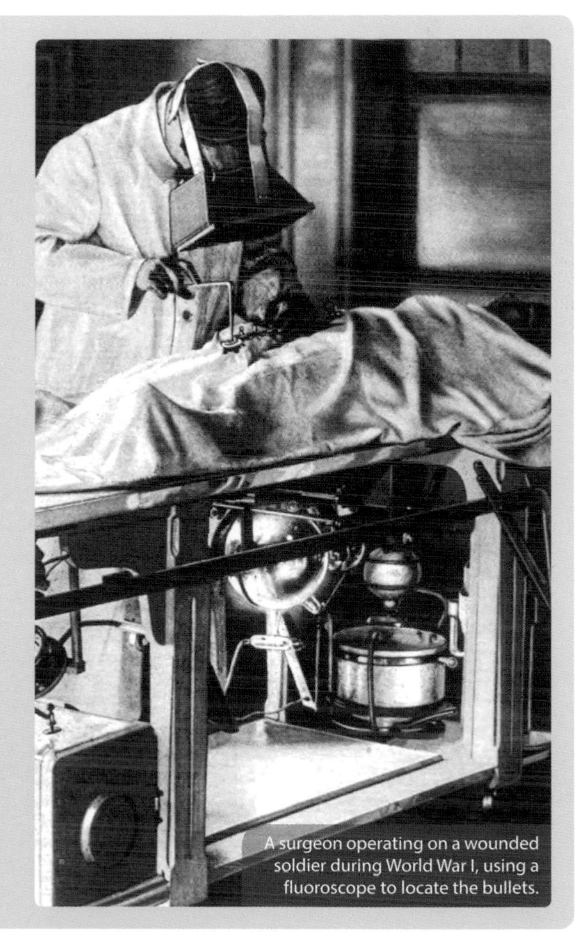

A surgeon operating on a wounded soldier during World War I, using a fluoroscope to locate the bullets.

MARY A. DELANEY

A LEADING DOG DEVICE
INVENTION: THE RETRACTABLE DOG LEASH

Not much is known about Mary A. Delaney apart from the fact that she was a New Yorker and that in 1908 she was the filer of the first dog leash patent in the United States. There were, as the patent stated "certain new and useful improvements" to the customary dog leash, whose drawbacks were outlined:

> *It is usually desirable that the dog should have a certain freedom in running about, but it is difficult to prevent the animal from running on the wrong side of lamp posts or pedestrians, thus causing much annoyance to the owner, who is constantly required to adjust the length of the leash in her hand and frequently the leash is dropped and the dog permitted to run away.*

To put an end to any such inconveniences the new leash had a drum and spring innovation that allowed the holder to let the leash out in stages. It was "particularly adaptable for ladies." The patent describes it:

> *In operation the leash is secured to the dog's collar and the handle is held in the hand. As the dog runs about and runs further away from his master or mistress, the leash is paid out and the spring is wound up. As the dog runs nearer his master or mistress, the spring automatically rotates the drum within the casing and winds up the chain or leash, thereby automatically preventing the leash from becoming entangled, and obviating the necessity of winding the leash about the owner's hand to take up slack.*

A modern retractable leash for a dog.

THE DOGS OF NEW YORK

The nineteenth century was a difficult time for dog owners. New York City, for instance, had laws about dogs having to be on leashes, but they were not strictly enforced for the 200,000 dogs with homes and many of them wandered the city streets. To make matters worse, there were estimated to be around 155,000 stray dogs in the city. There were numerous scary headlines about the actions of these dogs and there was also the risk of rabies which frightened people into demanding that dogs be muzzled with a shoot-to-kill policy for strays.

Vigilantes took dogs off the streets for a bounty of 50 cents an animal which was supplemented from 1850 by the establishment of a Dog Bureau whose employees were authorized to club dogs to death. The dog-catchers were viewed as thuggish and brutal for good reason. It made little difference, however, and the *New York Daily News* complained in 1856 that dogs "swarm in all the streets, obstruct the pavements, make night hideous with their howls and have a worse name than Aldermen in New York."

At the same time, Victorian attitudes to dogs were changing and the notion of the pampered middle-class pet began to emerge. Dogs were assimilated into family life and pet shops began to open. The owners of these animals were horrified by the activities of the dog catchers. Finally, by 1894, the American Society for Prevention of Cruelty to Animals had taken over dog-catching responsibilities from agents appointed by the city and bounty hunters.

MARION DONOVAN

ANOTHER FINE MESS

INVENTION: THE WATERPROOF DISPOSABLE DIAPER

Invention ran in the family of Marion Donovan (1917 – 1998), her father and uncle having invented a successful industrial lathe. Born in Fort Wayne, Indiana, as a child she spent time at her father's manufacturing plant. She graduated in 1939 from Rosemont College in Pennsylvania with a degree in English before being awarded a master's degree in architecture from Yale University, one of only three women on her course. She went into journalism, editing both *Harper's Bazaar* and *Vogue* magazines but gave it all up to stay home and look after her two children from her two marriages.

THE BOATER

By the time World War II came to an end, Marion, then living in Connecticut, had washed perhaps one diaper too many. Also tired of washing soiled bedding and clothing, she resolved to design a diaper cover that would keep her baby and the things around it dry. She took an old shower curtain and, seated at her sewing machine, created a waterproof cover that would fit over her child's diaper.

There were already rubber baby pants that were designed to perform the same function but they caused diaper rash and could pinch the skin of the child wearing them. Marion's cover was comfortable and did not cause a rash. She made slight improvements, replacing the dangerous safety pins that were normally used on diapers with snap fasteners and when it was finished, christened it the "Boater" because she thought it looked like a boat.

SELLING FOR A MILLION DOLLARS

In 1949, Marion presented her invention to manufacturers but as no one had the foresight to take the "Boater" on, she had to launch it herself, first putting it on sale at the department store, Saks Fifth Avenue. It was an immediate success. Within a couple of years, she sold her company and her patents to the Keko Corporation for $1 million.

She next worked on a diaper that would be fully disposable, and not just a cover. To do this, she had to find or make a type of paper that was strong but also absorbent. It had to be made so that the liquid was not in contact with the baby's skin. When she had finished it, she took it to diaper manufacturers and once again, no one was interested in helping her get it to market.

PAMPERS AND BEYOND

Ten years later, in 1961, a man named Victor Mills created a disposable diaper called Pampers that was inspired by Marion Donovan's work. Pampers became a national brand during the ensuing decades.

But Marion Donovan did not rest there. She continued to experiment with and invent things that had nothing to do with diapers, eventually owning twenty patents for creations such as a facial tissues box, a storage container box, a towel dispenser, a closet organizer and a hosiery clamp. She died at age 81 in a hospital in Manhattan, New York City.

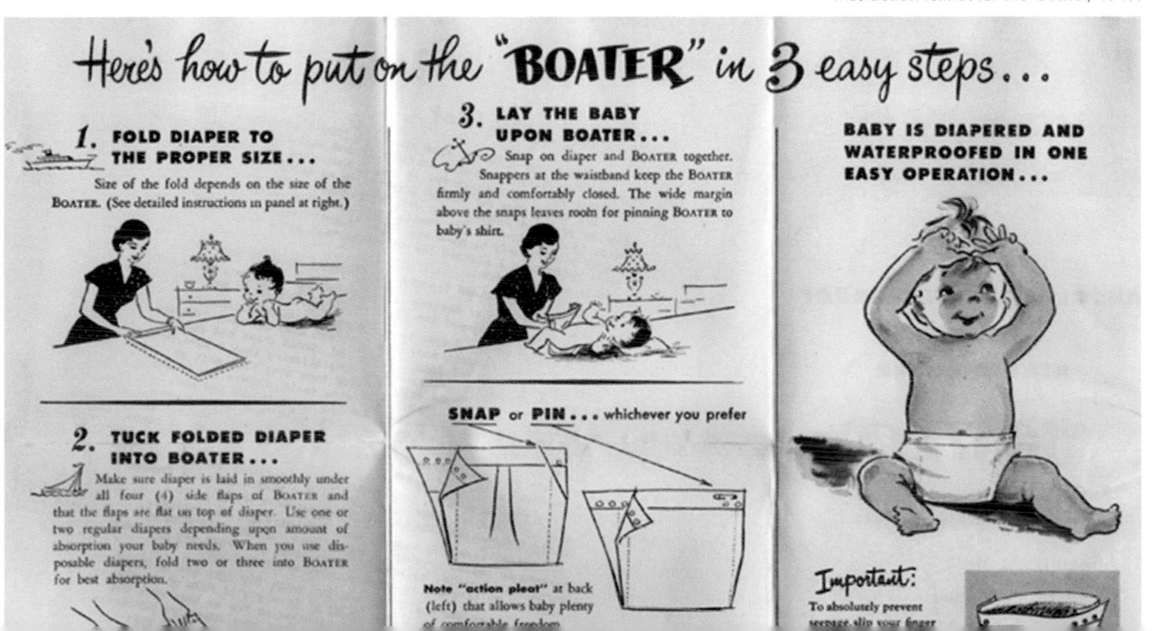

Instruction leaflet for the "Boater," 1949.

EMILY C. DUNCAN

COUNTING THE COST
INVENTION: THE BANKING CALCULATOR

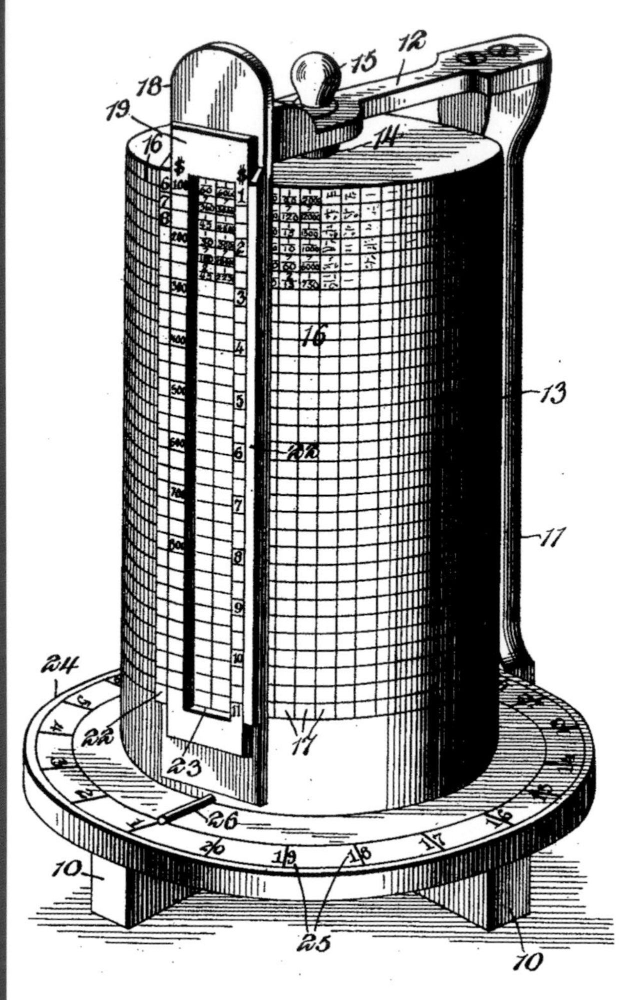

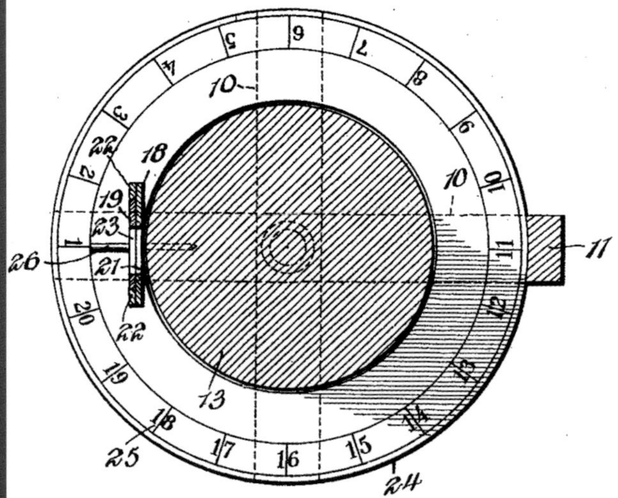

Patent diagram for Emily C. Duncan's calculator, November 24, 1903.

Emily Duncan invented banking calculators at the turn of the twentieth century, in the days before computers. She was born in 1849 in Coral, Illinois, as Emily Forbes. At the age of 14, she married her stepbrother, James Eugene Duncan who was a carpenter. He was 17 or 18 and home on leave from fighting in the American Civil War. The couple had eight children of whom five survived, living in Wisconsin and then on the Dakota frontier where life was tough. Eventually, the frontier proved too difficult and they moved back to Wisconsin.

A bevel square was Emily's first patent obtained at the age of 52. By this time, the family was living in Jennings, Louisiana, and it was while living here that she received her two patents for calculators. Her husband had already been granted a patent for a calculator, but hers were designed specifically for calculations involving credit or borrowing.

In her first calculator patent application, made in 1903, she stated that she wanted to be able to provide a "simple and readily understood structure with which computations ordinarily requiring considerable time and care may be accurately and quickly made." She added that her device should be of such a nature that "parts may be readily substituted, so that computations of different kinds may be made." This first device was made to compute interest at six, seven, or eight percent.

She was granted a further patent in 1904 for improvements in calculators. This device was designed to determine the number of days between any two dates within a given year to work out how much time was left on a loan. Unfortunately, she does not seem to have made much money from her inventions and she died in 1935 at age 85.

ELLEN EGLIN

WRINGING THE CHANGES

INVENTION: A CLOTHES-WRINGER FOR WASHING MACHINES

African American housekeeper Ellen Eglin, who was born in 1849 and lived in Washington DC, invented a clothes-wringer to be used with washing machines. Her wringer had two rollers located within a frame that was connected to a hand-crank. Clothes could be fed between the two rollers with one hand while the other cranked the rollers, squeezing all the water out of them. It was a wonderful labor-saving device at a time when washing was all done by hand.

Commercial "washing machines" or "wringers" had existed in England since around 1861, but there was nothing quite like Ellen Eglin's invention in the United States. Eglin sold her patent for just $18 to "a white person interested in manufacturing the product" because she feared that her color would deter white women from buying it. "You know I am black," she told the April 1890 issue of *Woman Inventor*, "and if it was known that a Negro woman patented the invention, white ladies would not buy the wringer. I was afraid to be known because of my color in having it introduced into the market, that is the only reason."

An example of a clothes-wringer that was used during the nineteenth century.

She explained to the magazine her keenness to have a patent issued in her own name:

> *I am working on another invention and have money to push it after the patent is issued to me, and the invention will be known as a black woman's. I am looking forward to exhibiting the model at the Women's International Industrial Inventors Congress to which woman are invited to participate regardless of color lines.*

Sadly, she never did manage to patent her second invention and we have no knowledge as to whether she did actually make it. She was next to be found working as a clerk in a census office and is believed to have died sometime after 1890.

AFRICAN AMERICAN WOMEN INVENTORS

An amazing number of African American women have made invaluable contributions to medicine, science, and technology, but often little is known about them apart from their inventions. This is partly, of course, because of the prevailing social attitudes before and after the abolition of slavery, but also because women were often prevented from patenting their inventions due to their gender.

Before 1865, slaves were also prohibited from registering patents and in the event that an enslaved person invented something, the patent was awarded instead to the slave owner. After the end of the American Civil War, African American inventors finally had the right to be granted patents. But African American women, in common with white women, still did not always receive the credit for their work and male family members were granted the patent instead. It is difficult, therefore, to know how many there actually were, and undoubtedly many have disappeared into the mists of time, uncredited for their ideas.

PRINCESS ELENA OF MONTENEGRO

CELEBRITY SELFIE

INVENTION: THE SIGNED PHOTOGRAPH

The British royal family and the King and Queen of Italy at Windsor Castle, England, 1904.

Princess Elena Petrović-Njegoš of Montenegro was the daughter of King Nicholas I of Montenegro and wife of King Victor Emmanuel III of Italy. Born in 1873, she was Queen of Italy from 1900 to 1946 and Queen Consort of the Albanians from 1939 until 1943. She was also the inventor of the signed photograph.

On December 28, 1908, the Sicilian city of Messina was struck by a massive earthquake and a tsunami in which the city was almost entirely destroyed and around 100,000 people lost their lives. Elena is shown in photographs of the time helping rescue workers to free and care for survivors which made her very popular among Italians.

During World War I, she worked as a nurse and turned the Quirinal Palace and Villa Margherita into hospitals for wounded soldiers. To raise funds for these initiatives she sold signed photographs of herself at charity desks, thus inventing the idea of the signed photograph now popular the world over with celebrities.

Queen Elena of Italy in a nurse's uniform during World War I.

Princess Elena of Montenegro, Queen of Italy, signed photograph, 1926.

GERTRUDE ELION

RATIONAL DRUG DESIGN

INVENTION: THE FIRST DRUGS FOR LEUKEMIA

Born in New York City, Gertrude Elion (1918 – 99) was an American biochemist and pharmacologist who had a Lithuanian father and a Polish mother. When she was seven, the family moved from Manhattan to the Bronx, where Gertrude indulged what she has described as an "insatiable thirst for knowledge." By the time it came to make some sort of decision about what path in life she was going to take, she found it difficult. Her beloved grandfather had died of cancer which made her decide to find a cure for that terrible disease.

MASTER OF SCIENCE

She enrolled at Hunter College in 1933, having decided to study science with the focus on chemistry. Fortunately, Hunter College provided free tuition because it was the time of the Depression and money was scarce. It also meant that she was unable to progress to graduate school, although she applied for jobs as an assistant and tried for scholarships. She taught biochemistry to nurses for a few months in the New York Hospital School of Nursing but luckily met a chemist who was searching for a laboratory assistant.

It was unpaid work, but she decided it would be useful to gain some experience. She did eventually save enough money to get into New York University in the fall of 1939 where she was the only woman in her chemistry class. Having to do research for her master's degree, Elion taught for several years, research filling her evenings and weekends. In 1941 she was awarded a Master of Science degree in chemistry.

WORKING WITH GEORGE HITCHINGS

A shortage of qualified chemists during World War II led to her working in a laboratory, but not in research. Instead, she found work in the quality control department of a food company. For an enquiring mind such as hers, the work was unfulfilling and repetitive. She wanted to continue learning and that was impossible in this role. Therefore, she found research work at

Gertrude Elion working in her laboratory.

George Hitchings and Gertrude Elion after winning the Nobel Prize for Medicine, 1988.

pharmaceutical manufacturer Johnson and Johnson in New Jersey, but was out of a job again when the laboratory shut down six months later.

There were a number of opportunities open to her in the world of research but she was intrigued by one position that was offered to her, as assistant to George Hitchings who worked at Wellcome Research Laboratories. This job took her from organic chemistry into microbiology, pharmacology, immunology, and virology. At the same time, she was pursuing her doctorate but eventually was told she could no longer study part-time. She made the critical decision to end her studies and continue with her work with Hitchings.

NEW DRUG TREATMENTS

Using a method known as "rational drug design," Elion and Hitchings were able to successfully interfere with cell growth, leading the way to a number of effective drugs for treating leukemia, gout, malaria, herpes, and many other illnesses.

In total Gertrude Elion developed forty-five patents in medicine. Among her inventions are 6-Mercaptopurine (Purinethol) which was the first successful treatment for leukemia and which is also used in the prevention of rejection in organ transplantation; Azathioprine (Imuran), the first immune-suppressive agent, also used in organ transplants; Allopurinol (Zyloric) which is used to combat gout; Pyrimethamine (Daraprim) which is used in the fight against malaria; Trimethoprim (Proloprim, Monprim and others) which is used against meningitis, septicemia and to fight bacterial infections of the urinary and respiratory tracts; Acyclovir (Zovirax) which is taken for viral herpes; and Nelarabine which is effective in the treatment of cancer.

NOBEL PRIZE

Gertrude Elion officially retired in 1983, but it made little difference to her as she continued her work at the lab, where she oversaw the adaptation of azidothymidine (known as AZT) which was the first drug introduced to combat AIDS. Her Nobel Prize in Physiology or Medicine was awarded in 1988 and was shared with George Hitchings and Scottish pharmacologist, Sir James Black. The citation stated that the award was given for "important new principles of drug treatment." In 1991 she became the first woman to be inducted into the National Inventors Hall of Fame. Gertrude Elion died at age 81 in 1999.

AMALIA ERIKSSON

SWIRLING DANCING CANDY

INVENTION: POLKAGRIS

Swedish candy sticks known as polkagris.

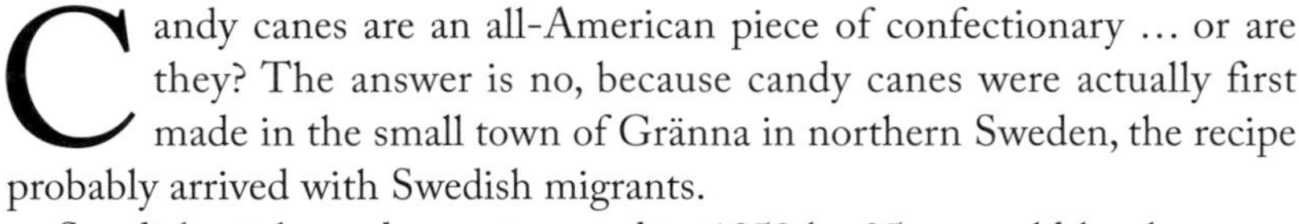

Candy canes are an all-American piece of confectionary … or are they? The answer is no, because candy canes were actually first made in the small town of Gränna in northern Sweden, the recipe probably arrived with Swedish migrants.

Swedish stick candy was invented in 1859 by 35-year-old local woman Amalia Eriksson (1824 – 1923), whose husband had recently died, leaving Amalia and her family in dire financial straits. She was given permission by the local town council to open a bakery where she planned to make and sell pastries and red and white peppermint stick candy, "polkagris" in Swedish.

The name came from the dance, the polka, which was all the rage in Europe when Amalia invented her product. The swirling nature of the dance is thought to have inspired her to name her confectionery after it. Her stick candy was made by twirling the red and white ribbons of sugar dough. "Gris" means "pig" in Swedish but at the time it was also slang for candy.

Amalia kept her recipe secret until she died, but it uses sugar dough that is boiled before being kneaded, pulled, and twisted until it is the requisite size. Peppermint and a small amount of vinegar are added. The original polkagris were straight, unlike the j-shaped American candy canes, and were red and white. Nowadays, they are produced in a range of colors, shapes, sizes, and even flavors including mojito, gin and tonic, strawberry and champagne, and bubble-gum.

Amelia died in 1923, but her polkagris is now the reason around a million tourists come to Gränna each year.

J-shaped American candy canes.

JUDITH ESSER-MITTAG

WORKING WITH THE FEMALE ANATOMY
INVENTION: THE DIGITAL-STYLE TAMPON

Judith Esser's o.b. tampons.

German gynecologist, Dr. Judith Esser-Mittag (born 1921, and usually known as Dr. Judith Esser) is the inventor of the digital tampon. "Digital" because it is applied using the fingers and does not need an applicator. Dr. Esser's aim was to create a product that worked with a woman's body to provide protection. Her profession as a gynecologist made her ideally suited to working out how to do that.

She had been dissatisfied, as a woman who enjoyed swimming, with what was available to menstruating women. The pads that were the most common method of providing menstrual hygiene could not be worn in water and she found the tampons that had to be applied with an applicator (invented by the American osteopathic physician, Earle Haas, in the 1930s) were uncomfortable and did not take women's bodies into account.

Judith Esser's o.b. tampon (from the German *"ohne binde"*—"without napkins") is much smaller than an applicator-tampon, and it is easy to insert, comfortable, and provides greater protection. Made of rolled fiber-pad layers it expanded in the same way all around, filling the vaginal cavity more effectively, ensuring there were no leaks. It was also more absorbent and worked with the female anatomy rather than against it.

The German Carl Hahn Company manufactured and sold Esser's tampon until the company was bought by Johnson and Johnson in 1974. Her woman-friendly idea is still widely used.

HISTORY OF THE TAMPON

For thousands of years, women have used tampons during menstruation. In fact, the world's oldest printed medical document, the Ebers Papyrus that dates to around 1550 BC, notes that Egyptian women were using tampons made of soft papyrus. Women of Ancient Rome used woollen tampons while in ancient Japan, women used paper tampons that were held in place by a piece of cloth. That would have been the business to be in, because the women of ancient Japan are reported to have changed their tampons ten or twelve times a day.

The first modern tampon was patented by Dr. Earle Haas in 1933. He said he got the idea from a friend who inserted a sponge into the vagina to absorb the menstrual flow. He created a plug of cotton that was inserted by means of a tube within a tube that meant the woman did not have to touch the cotton.

No one was interested, however, and later that year he sold the patent for $32,000 to a Denver businesswoman, Gertrude Tendrich. She named it Tampax and made the tampons at home using a sewing machine and a compression machine she obtained from Earle Haas. They first went on sale in 1936 and, now owned by Proctor & Gamble, Tampax is sold in over 100 countries.

FIONA FAIRHURST

HYDRODYNAMIC HUMANS
INVENTION: THE SPEEDO FASTSKIN SWIMSUIT

The European Patent office describes Fiona Fairhurst's invention as "a revolutionary swimsuit." She is a sports clothing designer who won the European Inventor of the Year in the Industry category in 2009 for her swimsuit. She had formerly worked at Speedo and had put many years into developing it.

The aim was to make a suit that would render the human body more "hydrodynamic," that, in other words, would help it to cut through the water with minimal resistance. A study was made of hydrodynamic animals which led to Fairhurst and her team focusing their attention on sharks. They wondered how the shark could be such an agile swimmer while creating a great deal of turbulence in the water. The answer lay, as they discovered, in the shark's skin which is covered with tiny ridges called denticles. These reduce the amount of water the shark's skin is coming into contact with as it swims. So Speedo produced a fabric that worked on the same principle as the shark's skin. The result was a knitted, water-repelling fabric that had printed denticles.

The Fastskin first appeared at the 2000 Olympic Games at Sydney and the results were immediate and astonishing. Some 83 percent of the medals were won by swimmers wearing the new swimsuit. Four years later, the second-generation Fastskin suit won a haul of forty-seven medals at the Athens Olympiad.

Underwater view of swimmers racing in Speedo Fastskin swimsuit.

MYRA JULIET FARRELL

DREAMING UP NEW INVENTIONS

INVENTION: DOMESTIC DEVICES, MEDICATION, AND MILITARY AIDS

Folding clothes line for compact storage in an apartment.

Myra Juliet Farrell (1878 – 1957) was an extraordinary woman who was born in Ireland. Her family immigrated to Australia while she was still a child, and she was brought up in Broken Hill, a mining city in New South Wales. Her parents were teachers and set up St. Peter's School where Myra was educated.

She began inventing at the age of just 10 when she devised a self-locking safety pin. She told a reporter that she often saw her inventions in dreams. She would scribble down what she had seen on the first thing that came to hand, which was sometimes the wall of the bedroom. This type of automatic writing done in her sleep, was always right to left which necessitated her reading it back in the morning with the use of a mirror.

The first of Myra's more than two dozen patents was for a device that enabled a skirt pattern to be copied directly onto fabric. This was followed by a folding clothes line for use in apartments; a boneless corset for sufferers of scoliosis; a brace for hernia sufferers; and a remarkable device that gave the wearer a facelift by mechanical means. Among her other inventions was a machine that picked and then packed fruit, a press stud that could be affixed to clothing without stitching, and the folding pram hood.

Incredibly, Farrell also invented medications, dreaming their formulae. She claimed to have dreamed the formula for an inhalation that reduced mucous and inflammation. She treated William Taylor, a young man suffering from tuberculosis with her inhalation and his condition underwent a dramatic improvement. The couple were married in 1906 and Taylor survived for a further six years. The product was marketed as Membrosus.

When World War I broke out, Farrell began work on a bulletproof and shellproof barricade and a light that could be projected over a great distance. Though the Australian Department of Defence showed some interest, it is unknown whether anything came of these ideas. A talented painter in what spare time she had between inventions, Myra Farrell was quite rightly described by one Australian newspaper as "a genius in the highest sense of the word."

SALLY FOX

GROWING COLORED COTTON

INVENTION: FOXFIBRE COTTON

Sally Fox inspecting bales of cotton.

Before Sally Fox invented Foxfibre cotton in the late 1980s, it was only possible to spin naturally colored cotton by hand, a long-winded process that persuaded businesses to choose white cotton instead and bleach it and dye it before spinning it on a machine. This provided people with the colored products they wanted, but it was a wasteful process that was bad for the environment, producing pollution as a consequence of the bleaching and dyeing processes.

Sally Fox had been working as a pollinator for a cotton grower, her job being to find pest-resistant plants. She started to breed brown and green cotton, selecting the seeds that produced the longest fibers and replanting them every year. The result was, after a good many years, two colored cottons that could be spun on a machine. In order to grow these and experiment further, she purchased a small plot of land and was granted Plant Variety Protection Certificates that are the equivalent of patents in the plant world.

By the early 1990s, Sally Fox was at the head of a $10 million a year business. Her company was producing naturally colored cotton for major clothing brands such as Levi Strauss, Land's End, and L.L. Bean. It has not all been plain sailing because many of the mills spinning cotton have moved to South America and Southeast Asia and her business has suffered, but she continues to make new naturally colored cottons, each one taking about ten years to come to fruition.

ROSALIND FRANKLIN

SOLVING SCIENCE'S GREAT MYSTERY
INVENTION: THE STRUCTURE OF DNA

It was not until the 1990s that British physical chemist Rosalind Franklin (1920 – 1958) began to be granted some of the recognition she was due for providing vital experimental data that allowed James Watson and Frances Crick to take the credit for discovering the structure of DNA in 1953, work that earned them the Nobel Prize for Physiology or Medicine in 1962.

In their Nobel Prize acceptance speeches they made no mention whatsoever of Franklin and until Watson's 1968 book *The Double Helix*, she was airbrushed out of existence. If she did get a mention, it was usually for stimulating their work rather than for providing the key evidence that supported their discovery.

RAISING THE STATUS OF HER SEX

Rosalind Franklin was born in London in 1920. Her father Ellis Franklin was a merchant banker, like generations of his family before him, and Rosalind grew up to be an independently minded young woman. She attended St. Paul's Girls' School where she demonstrated an early interest in physics and chemistry. At Newnham College, Cambridge, she studied natural sciences with a particular focus on chemistry, graduating in 1941.

Her degree qualified her for a graduate research scholarship, but her supervisor objected to her "raising the status of her sex to equality with men." This led her to move into war-related work for the British Coal Utilization Research Association, publishing a number of papers during that time which earned her a PhD from Cambridge University in 1945.

She moved to Paris to work with the crystallographer, Jacques Mering, at the Laboratoire Central des Services Chimique de l'État where she learned about x-ray diffraction which was critical in the research that she later undertook that led to the discovery of the structure of DNA.

PHOTOGRAPH 51

In January 1951, back in England, Franklin began working as a research associate in the biophysics unit at King's College London. She was encouraged there to employ her expertise in x-ray diffraction techniques on DNA fibers. She and the student with whom she was working, Raymond Gosling, made an astonishing discovery. When they photographed DNA, they discovered that it actually had two forms, a dry "A" form and a wet "B" form.

One of their x-ray diffraction images which is now known enigmatically as "Photograph 51" provided critical evidence in the identification of the structure of DNA. Photograph 51 was obtained following hundreds of hours of x-ray exposure using a machine that had been created by Rosalind Franklin.

MY PULSE BEGAN TO RACE

But in 1953, Franklin's colleague Maurice Wilkins with whom she did not see eye to eye, showed Photograph 51 to competing scientist James Watson without permission. Watson, who was also working on DNA with Francis Crick at Cambridge, has since admitted that when he was shown the image

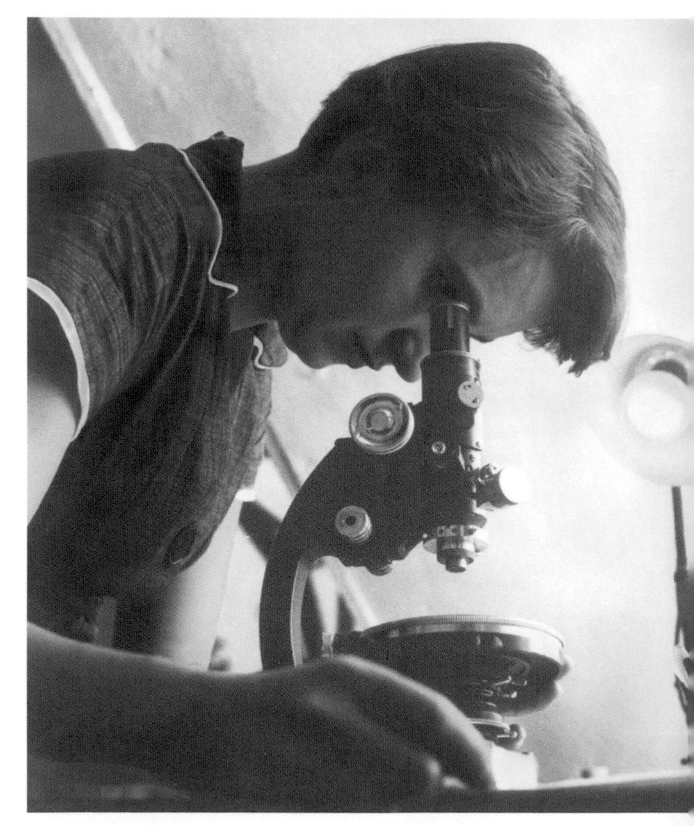

Rosalind Franklin in a laboratory looking into a microscope.

"My jaw fell open and my pulse began to race."

Watson and Crick used Rosalind Franklin's photograph as the basis for their famous model of the structure of DNA which was published on March 7, 1953, the Nobel Prize followed nine years later. Their model was published in *Nature* magazine in April 1953.

In a footnote they acknowledged Franklin, saying that they had "been stimulated by a general knowledge" of the work being done by Franklin and Wilkins. The reality was that much of their work had its foundations in Franklin's photo and her research.

OUTSTANDING CONTRIBUTION

Franklin never complained, maintaining the stiff upper lip that her upbringing had taught her. She left King's College in March 1953, taking a position at Birkbeck College and published a number of papers on viruses and structural virology.

Rosalind Franklin, a partner in solving one of science's great mysteries, died of ovarian cancer in 1958 at age 37, long before her outstanding contribution was ever admitted by the Nobel Prize winners Watson and Crick, or recognized by the world of science.

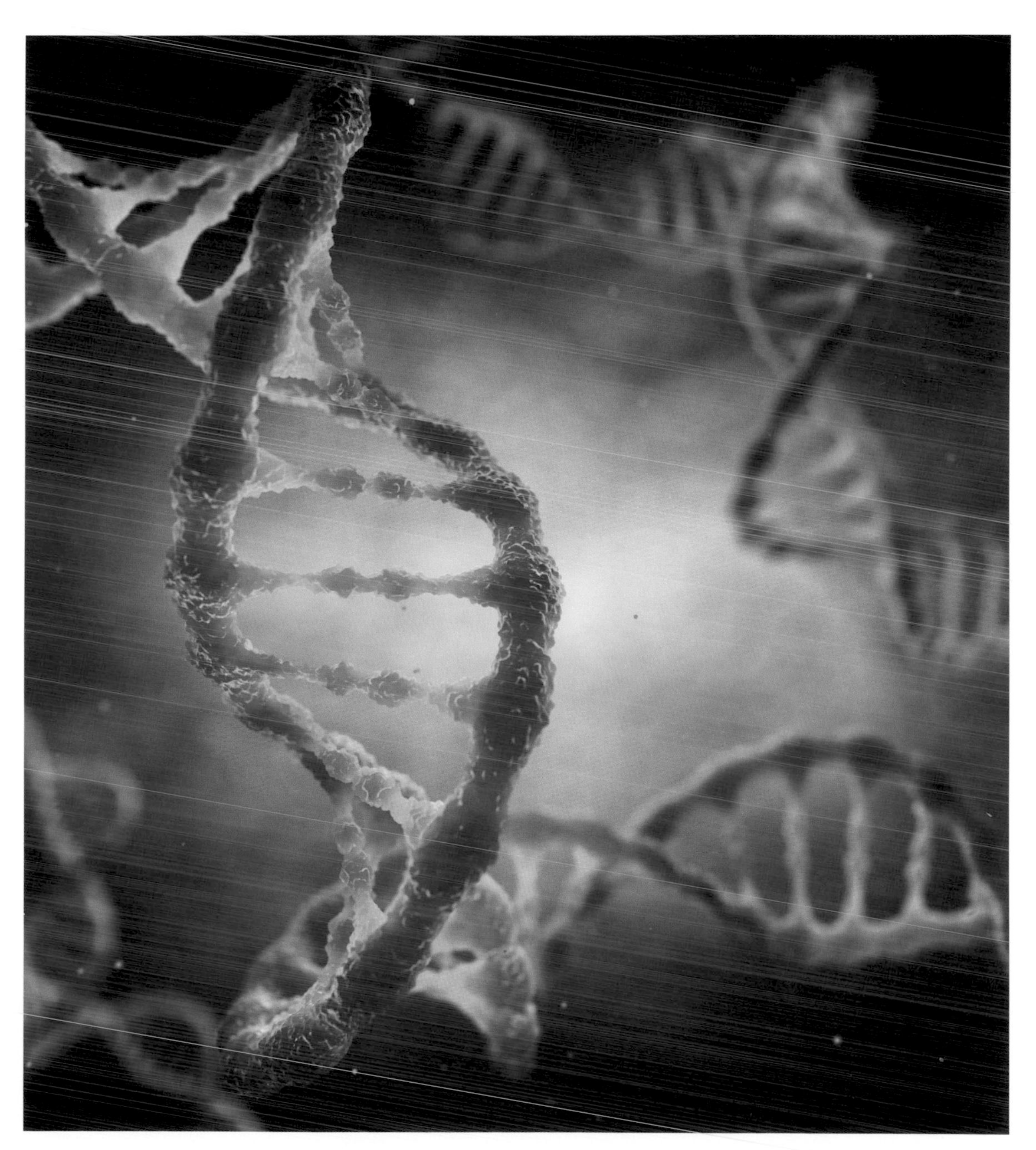

The DNA double helix molecule.

HELEN MURRAY FREE

CHANGING MEDICAL DIAGNOSTICS
INVENTION: SELF-TESTING SYSTEMS FOR DIABETES

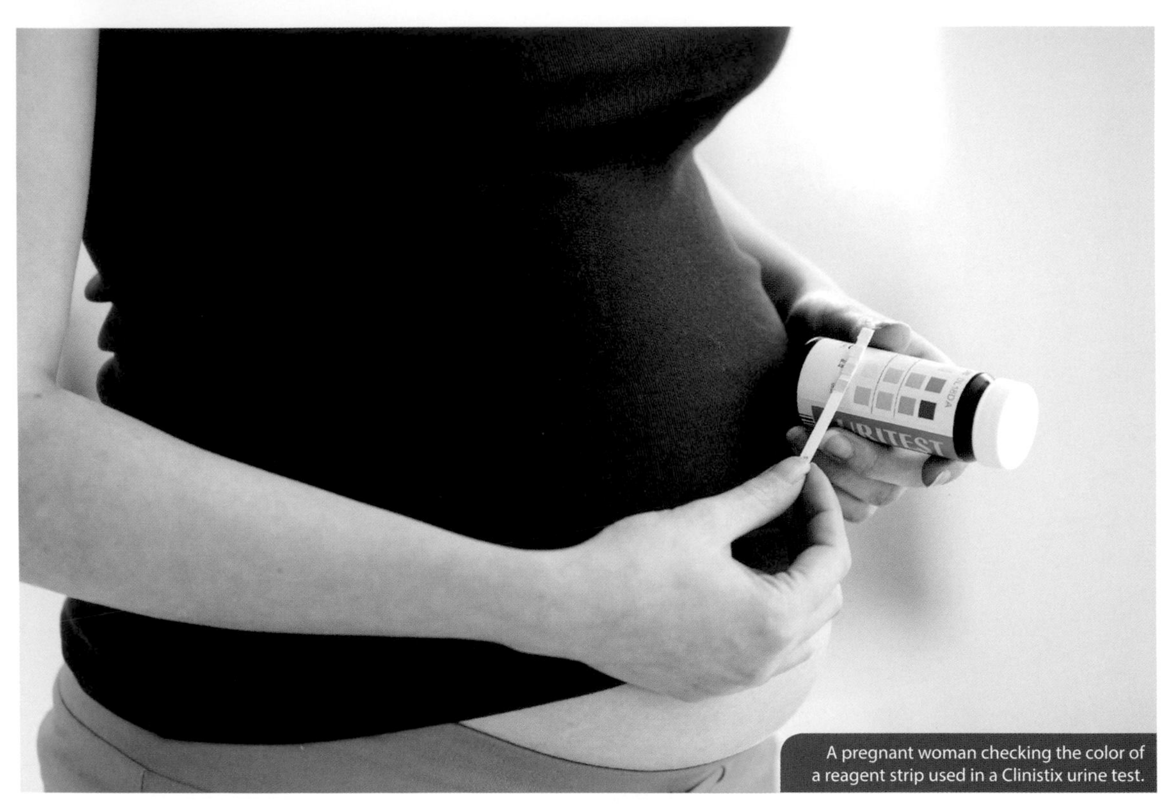

A pregnant woman checking the color of a reagent strip used in a Clinistix urine test.

Born in Pittsburgh, Pennsylvania, in 1923, Helen Murray was the daughter of a coal company salesman whose wife died when Helen was just 6 years old. At the College of Wooster in Ohio, her aim was to study English and Latin but it was around this time that the Japanese launched their surprise attack on Pearl Harbor, taking the United States into World War II. Men were being drafted into the US Army, and there was a need for women to pursue careers in science. She switched courses to chemistry and graduated in 1945.

After graduating, she went to work, finding a job as a quality control chemist at Miles Laboratories (where Alka-Seltzer was developed) which was later acquired by the chemical giant, Bayer Diagnostics. Her heart lay in research, however, and when the noted biochemist Alfred Free, working for the same company, offered her a position in his research team, she grabbed it. She married Free in 1947 and the couple went on to have six children. Not only did they become partners in life, they also created an important scientific partnership that would change the nature of medical diagnostics.

Their early work focused on improving the Clinitest. This was a tablet that had been developed to measure the amount of glucose in the urine of people suffering from diabetes. Their research led to Acetest which was another test for diabetes involving a tablet. The culmination of their work in this area was Clinistix, a chemically coated dip-and-read stick which changed color when dipped in a urine sample. The intensity of color change indicates the amount of glucose present. They extended this later to enable doctors to test levels of key indicators in other diseases and in 1975 they published *Urinalysis in Laboratory Practice*, which became a standard work in their discipline.

Helen Free retired in 1982, but her dip-and-read strips are now widely available at low cost and are used not just for diabetes but also for pregnancy and numerous other conditions, saving countless numbers of lives.

IDA FREUND

SHAPING THE TEACHING OF SCIENCE
INVENTION: PERIODIC TABLE CUPCAKES AND A GAS MEASURING TUBE

Periodic Table Cupcakes at the Chemical Heritage Foundation, Philadelphia, PA, USA.

Ida Freund (1863 – 1914) had a few firsts to her name. She was, for instance, the inventor of Periodic Table Cupcakes. Each cupcake represented an element, its name and atomic number shown in icing when the cakes were laid out in the periodic table format. Nowadays, this culinary innovation is a means of celebrating chemistry at school fairs and parties and helps promote public engagement with science. More significantly, however, she was also the first woman chemistry lecturer in the UK and was a big influence on the teaching of science to women.

Born in Austria, she studied in Vienna before moving to England in 1881. Her uncle, violinist Ludwig Straus, got her a place at Cambridge University where, even though English was her second language, she earned a First Class Honors degree in chemistry.

She spent a year teaching at the Cambridge Training College before becoming a demonstrator at Newnham College, Cambridge. Finally, in 1890, she was appointed as a lecturer in chemistry, a position she retained until 1913. Soon after becoming the first woman chemistry lecturer, however, she underwent surgery on her leg, most of which she had lost in a cycling accident as a girl.

In 1893, she returned to work at Cambridge in a wheelchair but that did not impede this very determined lady in any way. She still worked and traveled and was an inspirational teacher as well as an active feminist who strongly advocated women's suffrage.

She helped in the struggle for women to be able to gain admission to the Chemical Society in the early twentieth century. She championed the teaching of science in girls' schools and as women were not permitted in the same laboratories as men at Cambridge, she held special classes at Newnham College.

Ida Freund was also a skilled laboratory chemist and a practical researcher as well as a writer of a couple of textbooks on chemistry. Her main invention was a gas measuring tube that was named after her, sadly no longer in use, but possibly her most memorable achievement was her influence on the way that science is taught.

FRANCES GABE

A TRUE AMERICAN ORIGINAL
INVENTION: THE SELF-CLEANING HOUSE

American artist and inventor, Frances Gabe (1915 – 2016) hated housework, which is why in 1984 she invented the self-cleaning house. Born Frances Grace Arnholz in Boise, Idaho, her mother died when she was very young, and so she spent her time with her father, a building contractor and architect. Construction workers were her family and they taught her about building.

At age 17, she married an electrical engineer, Herbert Bateson. But he never had much work, forcing her to find a means of feeding her two children. She launched a home repair business in Portland and even though she made a success of it, her work-shy husband was so embarrassed that he insisted she stop using his name.

She decided, therefore, to be known by her initials "GAB," taken from her name Grace Arnholz Bateson with a lower case "e" appended at the end. Thus, her name became Frances Gabe. Not long after, she and her husband called it a day and divorced.

Her idea for a self-cleaning house began to develop around 1955, as a means of avoiding housework. During the next twenty-five years she was awarded more than sixty patents. The structure she built was 30 feet by 45 feet and its main cleaning device was a spray with two nozzles. It sprayed a fine mist of soap which was immediately followed by a rinse with water and drying with hot air. All the floors sloped slightly toward a drain.

The house also featured a closet that washed and dried clothes; a kitchen cupboard that was also a dishwasher; drawers that allowed dust to fall to the floor; a dry toilet; and a fireplace that hosed away its own ashes. Furniture was waterproof and there were no carpets or curtains.

She later explained that she designed the house as she realized that one day she would be too old to clean the place. Frances Gabe died in obscurity in 2016 at age 101. A once-celebrated inventor, she was the creator, and for a long time the sole inhabitant, of the world's only self-cleaning house.

Frances Gabe made her self-cleaning house do its own scrubbing.

PRATIBHA L. GAI

CREATING MORE FUNDAMENTAL SCIENCE INVENTION: THE ATOMIC-RESOLUTION ETEM

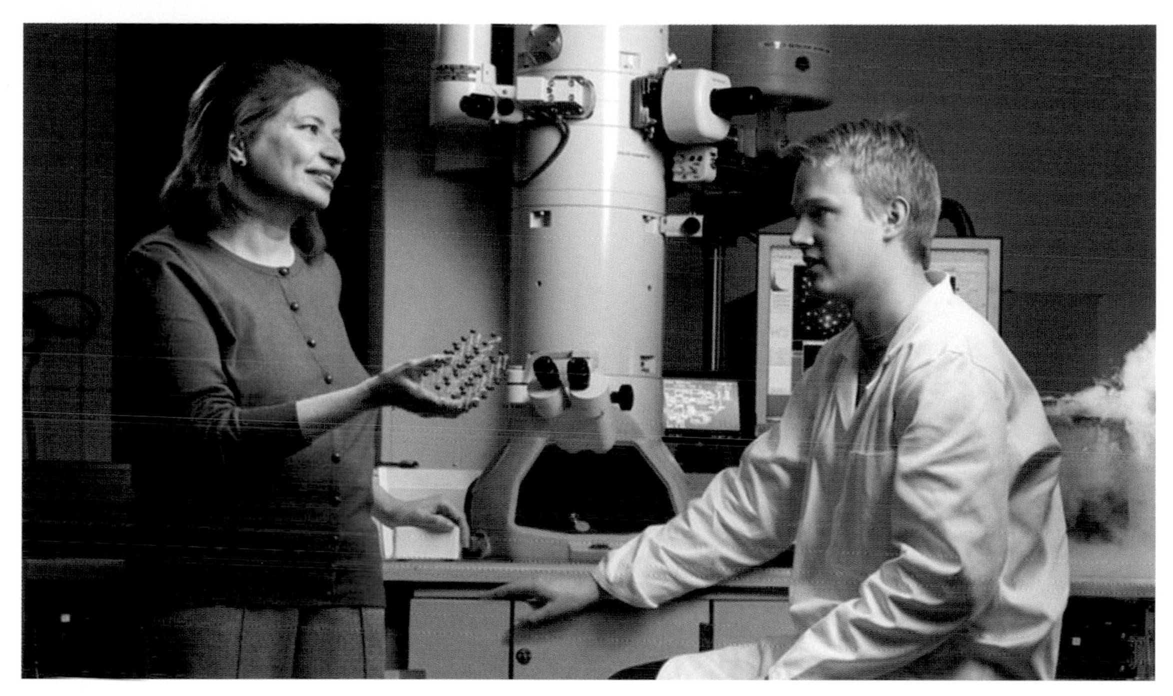

Pratibha Gai with her ETEM which can observe elements even smaller than those seen with an electron microscope.

It took twenty years for Pratibha Gai to build her electron microscope, constantly refining and improving it. The work began when she was a postdoctoral researcher at Oxford University and went on for the next eighteen years at the US chemical firm DuPont and at the University of Delaware. Her Environmental Transmission Electron Microscope (ETEM), which was completed in 2009, is capable of visualizing and analyzing at the atomic level dynamic gas-catalyst reactions that underpin chemical processes.

Yet it was very difficult for her to get into science in the first place. She had grown up in India with a fascination for the subject. But it was not socially acceptable in her society for a woman to pursue a career in the physical sciences. Nonetheless, she won a scholarship as a result of being selected as a national science talent and went to study at Cambridge University. In 1974, she was awarded a PhD for the research she had undertaken on weak beam electron microscopy.

Gai decided not to patent her invention even though it could have proved extremely lucrative for her. "I thought," she has said, "that if I patented it, no one else would be able to work with it. I might earn some money, but I was not interested in that. I was interested in applications for many researchers, creating more fundamental science. So, I decided not to patent it." Pratibha Gai is now Professor and Chair of Electron Microscopy at the Departments of Chemistry and Physics, University of York, England, and is a very vocal advocate for women with careers in science.

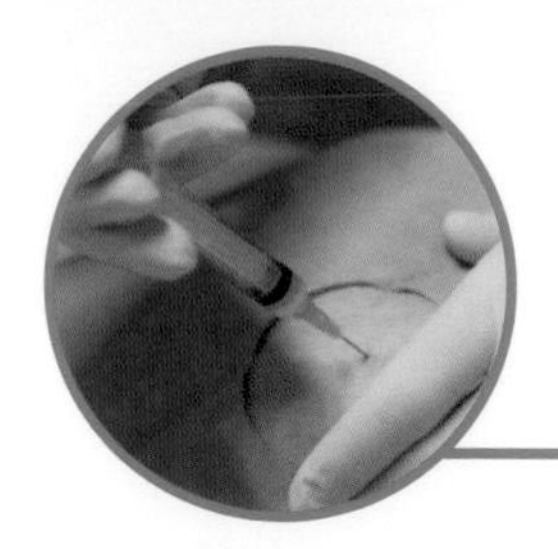

LETITIA MUMFORD GEER

SAFE SELF-INJECTION
INVENTION: THE ONE-HANDED SYRINGE

Before the American Letitia Mumford Geer's invention, syringes had to be operated using two hands. That changed in 1899, when she was granted a patent for what was described as an "In a hand-syringe." It consisted "of a cylinder, a piston and an operating-rod which is bent upon itself to form a smooth and rigid arm terminating in a handle, which, in its extreme positions, is located within reach of the fingers of the hand which holds the cylinder, thus permitting one hand to hold and operate the syringe." The patent describes the operation of the syringe:

> *The handle can be drawn into a position near to the cylinder while injecting the medicine by the use of one hand, thereby enabling the operator to use the syringe himself without the aid of an assistant. The advantages of the medical syringe are several. The syringe is very simple and cheap. It can be operated with one hand.*

Geer's invention was one of those that undoubtedly saved a lot of lives. Sadly, nothing much else is known about her.

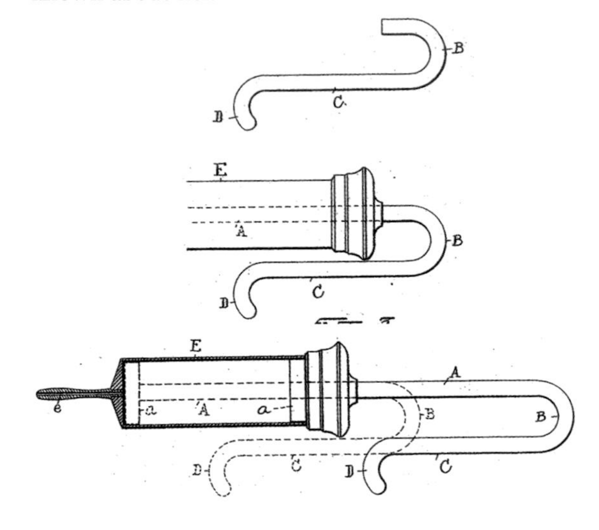

Patent diagram of Letitia Mumford Geer's syringe, April 11, 1899.

THE HISTORY OF THE SYRINGE

We know from the medical work, *De Medicina*, an Ancient Roman medical Encyclopedia written by Aulus Cornelius Celsus who lived between around 25 BC and 50 AD, that piston syringes were being used at that time to treat medical complications. A significant development happened in the ninth century AD, however, when the Egyptian-Iraqi surgeon Ammar ibn Ali al-Mawsili invented the "injection syringe," a needle used for the extraction by suction of soft cataracts. It is thought that the syringe was used in this way until the thirteenth century.

In the seventeenth century, as a result of his investigation of the scientific principles on which hydraulics worked, French mathematician Blaise Pascal invented a new type of syringe. It was used to test his theory that pressure exerted anywhere in a confined fluid is transmitted equally in all directions and that the pressure variations remain the same.

The hollow needle was invented by an Irishman, Francis Rynd who used it to make the first subcutaneous (beneath the skin) injections in 1844. Nine years later, French doctor Charles Pravaz and Scottish doctor Alexander Wood developed a medical hypodermic syringe that had a needle fine enough to pierce the skin. Wood tried some experiments with injected morphine in the treatment of nervous conditions and sadly, he and his wife became addicted to morphine. Mrs. Wood became the first victim of an injected drug overdose.

Letitia Mumford Geer's one-handed syringe arrived in 1899 and in 1946 the first all-glass syringe with an interchangeable barrel and plunger was made by Chance Brothers in England. The revolutionary nature of this cannot be overestimated. It permitted all the different components to be mass-sterilized without having to match them all up afterwards.

The world's first plastic, disposable syringe, made from polyethylene was created by Australian inventor Charles Rothauser in 1949. Polyethylene softens with heat which meant that the syringes had to be chemically sterilized before packing. This made them expensive. He produced the first injection-molded syringes a couple of years later, made of polypropylene, a plastic that can withstand heat. Many different disposable syringes followed and these are now commonplace in medicine.

MARY HOPKINSON GIBBON

INVENTING THE HEART-LUNG MACHINE
INVENTION: A PUMP-OXYGENATOR

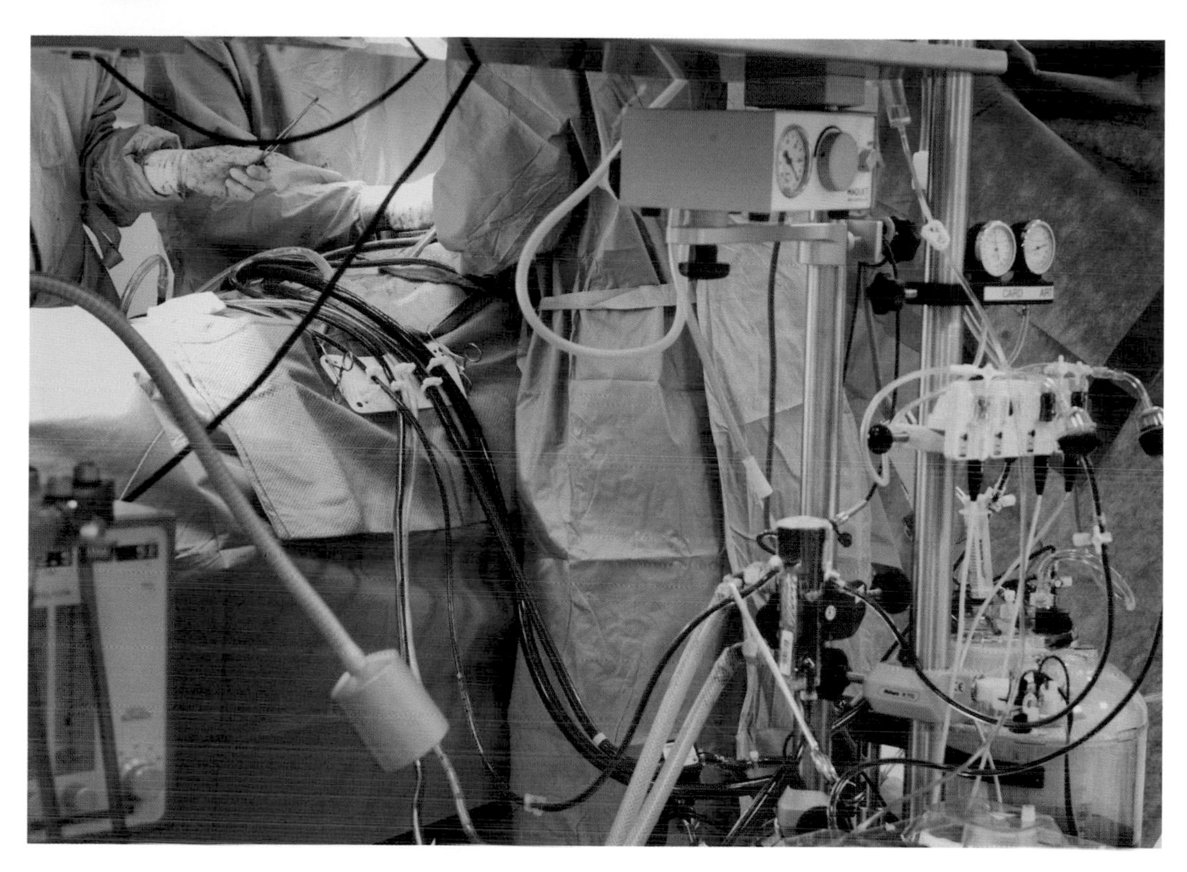

A cardiopulmonary bypass, or heart-lung machine, has taken over the heart and lung functions of the patient during surgery.

Mary Hopkinson Gibbon's pump-oxygenator is one of the most significant inventions in recent medical history. It has made it possible for open-heart surgery, heart and heart-lung transplants, and for lengthy surgical procedures to be carried out.

Mary Gibbon (1905 – 86) was working at Massachusetts General Hospital with her husband, surgeon John Heysham Gibbon, and as early as 1935 foresaw the prospect of open-heart surgery. The idea of the pump-oxygenator that she devised with her husband was to temporarily do the work of the heart and lungs while the surgery was taking place.

It served as a prototype for the first successful heart-lung machine that in 1953 took over the heart and lung functions for 26 minutes during an operation on a teenage girl in which John Gibbon repaired a hole in her heart. Gibbon's subsequent open heart surgeries revolutionized cardiac surgery in the twentieth century.

John Gibbon and his wife Mary Hopkinson developing the first heart-lung machine.

BARBARA GILMOUR

RUMORS OF BEWITCHING

INVENTION: DUNLOP CHEESE

After final pressing the cheeses are placed in temperature-controlled cheese stores where they will mature for many months, at the Dunlop Dairy, Ayrshire, Scotland.

It was while she was in exile in Ireland from Scotland around 1660 because of her Presbyterian beliefs, that Barbara Gilmour learned about the art of making whole milk cheese. Her exact origins are unknown, but her surname was a common one around the "Lands of Chapeltoun" in East Ayrshire. She ended her exile after the "Glorious Revolution" of 1688 when King James II of England was overthrown and replaced on the British throne by William III, Prince of Orange. Back in Scotland, she settled in Dunlop, East Ayrshire.

She had brought back with her the cheese recipe she had learned in Ireland. She proceeded to combine the best of Scottish and Irish cheesemaking methods to produce Dunlop cheese. Her cheese, made from unskimmed milk from Ayrshire cows, was at the time unknown in Scotland but it quickly became very popular and was copied by other farmers.

Until then, Scottish cheese had been of inferior quality, the cream being skimmed off the milk to make butter. There was some controversy about the cheese she made, the local people believing it to be impossible to make cheese from whole milk. Rumors began to spread that Barbara must be a witch—a dangerous accusation in those days that could have resulted in her being burned to death.

Nonetheless, she traveled the country, teaching how to make Dunlop cheese which became popular all over Scotland and further afield. The eighteenth and nineteenth century English writer William Cobbett described it as "equal in quality to any cheese from Cheshire, Gloucestershire, or Wiltshire."

Barbara Gilmour died in 1732 and Dunlop cheese finally stopped production in the area around 1940 although it is still available elsewhere.

SARAH E. GOODE

THE SPACE WOMAN
INVENTION: A FOLDING CABINET BED

BOYINGTON'S
PATENT AUTOMATIC CABINET FOLDING BEDS.
MOST COMPLETE AND PERFECT ARTICLE OF THE KIND NOW MANUFACTURED
Made in Bureau, Bookcase, Dressing Case, Sideboard, Cabinet, and Writing Desk styles.
NEW YORK WAREROOMS, 88 Mulberry St., near Canal, New York City. E. H. NORTON, MANAGER.
OF ICE AND FACTORY, 1463, 1465, & 1467 State St., Chicago, Ill. L. C. BOYINGTON, Pat. and Sole Mfr.

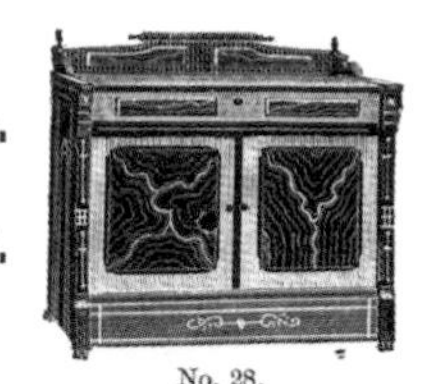
No. 28.

No. 22.

No. 34.

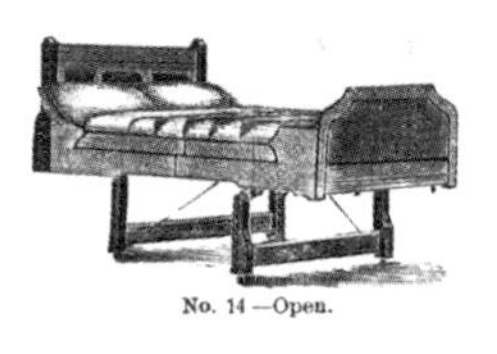
No. 14—Open.

No 36.

No. 10.

Advertisement for Automatic Cabinet Folding Beds,1881.

For some reason, Sarah Elisabeth Jacobs (1855 1905) as she was born, used to tell people she was born in Spain, even though her birthplace was Toledo, Ohio. Perhaps it was to hide the fact that she was the child of parents of mixed race, Oliver and Harriet Jacobs. At the end of the American Civil War, the family moved to Chicago and it was there that Sarah met her husband, Archibald Goode. They had six children, only three of whom survived to become adults. Her husband worked as a "stair builder" and upholsterer while she opened a furniture store.

Sarah Goode's invention came from a need that most people had at the time. Homes or living quarters were often very small and space was, therefore, at a premium. She learned this first-hand from customers at her store who constantly complained about how little space they had in their homes for the furniture she was selling. To service this need, she dreamed up an ingenious folding cabinet bed that made good use of the available space.

Folded up, it resembled an elegant roll-top desk with room for storage in it. But not only did it resemble a desk, it fulfilled this function perfectly, and at night it was unfolded to make a comfortable bed. She was granted a patent for her folding cabinet bed in July 1885, the first African American woman ever to receive a United States patent.

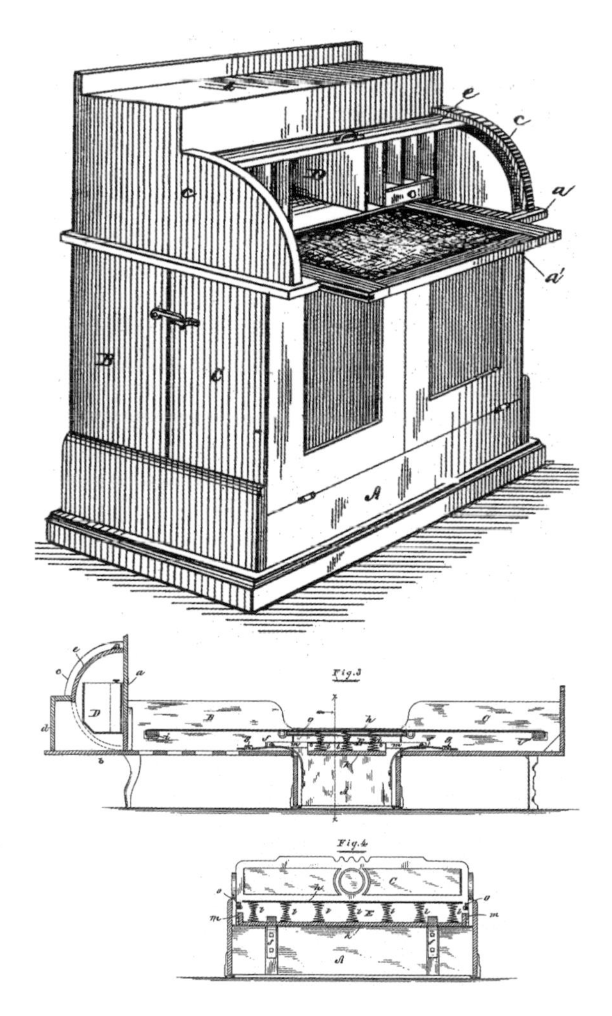

Patent diagram by Sarah E. Goode for a cabinet bed, July 14, 1885.

BETTE NESMITH GRAHAM

PAINTING OUT THE ERRORS
INVENTION: LIQUID PAPER

Using correction fluid to fix poor spelling.

Born in Dallas, Texas, in 1924, Bette Clair McMurray dropped out of high school at 17 and enrolled at secretarial school. Bette married Warren Nesmith just before he enlisted for World War II but the marriage broke up after the war and they divorced in 1946. When her father died in the early 1950s, Bette, her mother, and baby boy Michael moved to a property her father had left to her in Dallas.

She found a secretarial job at Texas Bank and Trust in Dallas. Electric typewriters were the cutting-edge office technology of the time but early machines were problematic. Their carbon-film ribbons made it almost impossible to neatly correct mistakes with an eraser, and entire documents had to be retyped from the beginning. Bette described how the idea occurred to her as she watched workmen painting a festive scene onto a window of the bank. She noticed how "an artist never corrects by erasing, but always paints over the error. So I decided to use what artists use. I put some white water-based tempera paint in a bottle and took a watercolor brush to the office. I used that to correct my mistakes."

For the next five years, Bette used this simple but brilliant method to correct her typing errors. Her correction fluid was named "paint out" by her co-workers who often came to borrow some. Eventually, she decided it might be worth trying to make some money from it and in 1956 it began to be sold under the brand name "Mistake Out."

It started to take off and when she got round to launching her own company, she renamed it "Liquid Paper." Initially, she worked from home, making her product in her kitchen, and then from a shed in her back garden. But Liquid Paper quickly became an essential item in every office.

In 1979, Bette Graham—she had married Robert Graham in 1962—sold the Liquid Paper brand to the Gillette Corporation for $47.5 million. Her 200 employees were producing and selling a staggering 25 million bottles of Liquid Paper a year. Six months after the sale of the company, Bette died suddenly at age 56.

BESSIE BLOUNT GRIFFIN

EATING WITHOUT HANDS

INVENTION: A FEEDING DEVICE FOR AMPUTEES

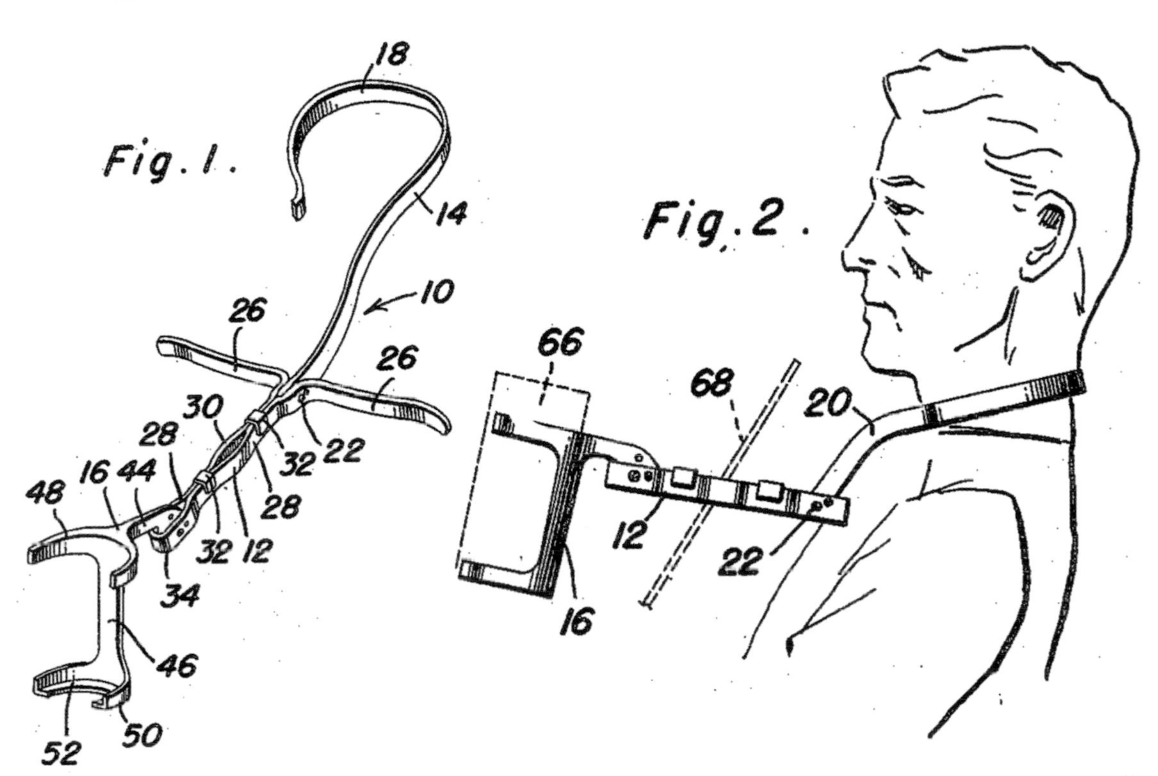

Patent diagram by Bessie Blount Griffin, April 24, 1951, for a portable receptacle support.

Bessie Blount was born in Hickory, Virginia, in 1914. She made important breakthroughs in assistive technologies as well as in her profession, forensic science, and has served as an important role model for African American women in their struggle for equality and recognition. Initially, she trained to become a physical therapist, studying in Chicago but after the end of World War II, she saw many soldiers returning from overseas as amputees. A large part of her work with such men was to help them find new ways to perform tasks without hands, feet, or legs, and one of the most difficult challenges for someone with no arms or hands is eating. The inability to perform this basic function often took away a person's independence and dramatically lowered their self-esteem.

Bessie devised an apparatus to allow amputees to feed themselves, a tube that delivered individual, bite-sized pieces of food to a mouthpiece. To have the next piece sent down the tube, he or she need only bite down on the tube. In 1951, Bessie developed a device that was much simpler. It used a neck-brace that provided support for a bowl, cup, or dish. Unfortunately, however, when she tried to interest the American Veterans Association in her devices, they failed to see the benefit they would bring to thousands of people. The only body that was really interested was the French government to whom she donated the rights to both inventions.

She continued to come up with clever inventions. A disposable cardboard emesis basin, made by molding and baking a mixture of flour, water, and newspaper was sold to Belgian hospitals, Americans once again showing no interest. In a variation of her design, it is still in use in Belgian hospitals to this day.

Bessie Blount changed careers in 1969, when she began working as a forensic scientist for police departments in New Jersey and Virginia, but that is another story in the life of this remarkable African American woman who died in 2009, at age 95.

SARAH GUPPY

BUILDING FOR THE FUTURE
INVENTION: PILING FOUNDATIONS FOR BRIDGES

Clifton Suspension Bridge spanning the Avon Gorge and River Avon, designed by Brunel and completed in 1864 in Bristol, UK.

Sarah Guppy was one of the foremost engineering inventors of the Georgian era in England. But her inventions had to be registered by her husband under the name "the Guppy family," as, at that time, women were not permitted to use their own names.

Born Sarah Beach in Birmingham, England, in 1770, she moved to Bristol when she married the wealthy Samuel Guppy whose family ran a sugar company. The couple had six children, and in 1811, she patented her first invention, a method of making safe piling foundations for bridges which paved the way for a new type of suspension bridge.

Her patent describes a system whereby vertical rows of tree trunks were driven deep into the ground and fixed together by a frame so that they would be able to resist erosion by water. The piers of the bridge were built on this base, and chains were then secured on top and pulled tight to create a platform on which a deck could be built.

Guppy was a friend of Isambard Kingdom Brunel, and her son worked for Great Western Railway which Brunel built. She constantly gave technical advice to the GWR directors, such as her recommendation to plant willow trees and poplars on the railway embankments to make them more stable and less liable to collapse.

There were ten "Guppy family" patents in the first half of the nineteenth century. These included a bed that had built-in exercise equipment and a dish that could keep toast warm. She also invented the fire hood, known as the Cook's Comforter, and a candlestick that was designed to make candles burn longer.

In 1837, at age 67, this inventive lady, now a widow, married a man 28 years younger than herself who, unfortunately, squandered her money on gambling. She died in 1852 at the age of 82.

MANDY HABERMAN

FEMALE INVENTOR OF THE YEAR
INVENTION: THE ANYWAYUP CUP

The Anywayup Cup was the world's first totally non-spill cup and has revolutionized the baby products market.

When the daughter of English inventor Mandy Haberman was born in 1980 with the genetic disorder Stickler Syndrome, Haberman went to work to try to improve the life of not only her own child but also other children with similar problems around the world. One of the manifestations of Stickler Syndrome is a cleft palate which makes it difficult to suck properly. So, Haberman devised a special bottle—the Haberman Feeder—that helps such children to drink properly. It was so effective that it is now used in hospitals not just in England but around the world.

Her next invention was a huge commercial success, being commended everywhere and winning many awards for her, including British Female Inventor of the Year in 2000. The Anywayup Cup prevents spillage when a child is drinking from it, even if it is turned upside down. It also protects growing teeth by permitting the liquid to flow only when a child is sucking and swallowing. Thus the child cannot engage in what is called "comfort sucking." Haberman's feeding innovations now cover a range of products.

MARY HALLOCK-GREENEWALT

THE ART OF PLAYING COLORS
INVENTION: VISUAL MUSIC

Mary Hallock-Greenewalt playing the Sarabet, an organ that played color instead of music

Mary Hallock-Greenewalt (1871 – 1950) was so far ahead of her time that even now her ideas are difficult to grasp. The lightshows at rock concerts and the flashing colors that sometimes accompany music on our various electronic devices all have their origins in her ideas which were generated back at the beginning of the twentieth century when she attempted to create an organ that played colors instead of music. It was called the Sarabet, after her mother, and she started work on it in 1906, fiddling with it and improving it until 1934.

Already an accomplished pianist, the Syrian-born Hallock-Greenewalt was trying to create an art form similar to music but for color. She called this *Nourathar* which is taken from the Arabic for the essence, flavor, or influence of light. Her color organ was designed to project a sequence of colored lighting that was arranged for special musical programs.

The creation of the Sarabet meant that she had to invent several new technologies and she was granted nine patents for these from the US Patents Office. One of these devices was a non-linear rheostat—a two-terminal variable resistor—the patent of which was infringed by a number of companies, including the mighty General Electric. Hallock-Greenewalt sued and won in 1932. She published a book, *Nourathar: The Fine Art of Light-Color Playing* in 1946 and died in 1950, at age 79.

HANNA HAMMARSTRÖM

WIRED FOR SOUND
INVENTION: TELEPHONE WIRES IN SWEDEN

Swedish telephone wires in winter.

Hanna Hammarström was born in 1829, the daughter of a cotton and silk merchant, who was determined that his daughters should work. To this end, he gave Hanna a machine that produced overspun metal wire. This was a product very much in demand with milliners who used it in ladies' hats.

When, in 1883, the Stockholms Allmänna Telefonaktiebolag (General Telephone Company) was launched, Hanna began to experiment with the production of telephone cable, employing the same techniques that she was already using for her hat wire. As ever, being a woman, she was not taken seriously but the founders of the phone company had confidence in her work and supported her. They found her premises in which to begin production and provided her with very cheap power for her machines. It would be a worthwhile venture for them if she was successful as up to that point all telephone wire had to be imported at great expense from Germany.

Her company had eight employees, all women whom she had trained. They operated five large gimping machines, spindle looms, and spooling machines, and as well as telephone wire they produced microphone wire, double-woven cable, and conduction cable. The determined and inventive Ms Hammarström worked in her factory until her death in 1909.

RUTH HANDLER

BE ANYTHING YOU WANT TO BE
INVENTION: THE BARBIE DOLL

By creating a doll with adult features, Mattel enabled girls to become anything they wanted.

The youngest of ten children of Polish immigrants, Ruth Mosko was born in Denver, Colorado. In 1938, she married her childhood sweetheart, Elliot Handler, and the couple moved to Los Angeles. Elliot made the furniture for their new house using two recently created plastics, Lucite and Plexiglas. This idea worked out so well that Ruth suggested Elliot go into business with it. They launched a furniture-making business with Ruth as the company's sales force.

Elliot soon launched another business, making picture frames, with Harold "Matt" Matson. They called the company "Mattel," a combination of their names Matt and Elliot. As time went on, they branched out into using leftover parts from the manufacturing process to create furniture for doll's houses which began to make more money than the picture frames and Mattel moved into solely producing toys.

Ruth had noticed that the paper dolls her daughter played with had a lot of limitations with paper clothes that did not fit. She proposed that Mattel produce a three-dimensional plastic doll and a wardrobe of clothes made of proper fabrics. They gave the doll an adult body with beautiful breasts as Ruth also had seen how her daughter projected her own adult future onto her dolls.

They named the doll "Barbie" and showed her at the 1959 New York Toy Fair. However, Barbie was not an immediate success. But when *The Mickey Mouse Club* made its debut on American television, Mattel advertised heavily, and featured Barbie. The doll then became an overnight sensation and quickly made them very rich. Soon, Barbie had a boyfriend, named Ken. More than a billion Barbies have now been sold in 150 countries.

Ruth and Elliot Handler ran Mattel for thirty years but were forced to resign in 1975 in the middle of a huge financial scandal. By this time, however, Ruth had already embarked on a second career, making prosthetic breasts for survivors of breast cancer. As she once joked, "I've lived my life from breast to breast." Ruth Handler died in 2002, at age 85.

MARIE HAREL

SECRETS OF THE CHEESE

INVENTION: CAMEMBERT WHEELS

A round of Camembert cheese.

There are a couple of versions of Marie Harel's invention of the soft, creamy Camembert cheese. In one, she is a 19-year-old cheese-maker, living in the village of Camembert who had fallen for the boy next door. As she was making cheese, her thoughts were of him and not cheese-making. She forgot to add the blue mold culture to one batch of cheese and the result was the cheese we now know as Camembert.

Another version suggests that Marie Harel actually deliberately grew the white mold *Penicillium candidum* on a well-known cheese as an experiment. She is rumored not to have been from Camembert at all but from Brie, another famous source of delicious creamy French cheese. Legend has it she came to Camembert in Normandy to marry a local farmer. On the couple's farm, she invented the small Camembert wheel cheeses.

Whatever we are to believe, Marie Harel undoubtedly was the inventor of Camembert cheese, as is stated on her gravestone in the Camembert churchyard and on a statue in the nearby town of Vimoutiers which is inscribed: "Marie Harel, 1791 – 1845. *Elle inventa le Camembert*. R.I.P." Her invention has turned into a phenomenal industry with half a million Camembert cheeses now being produced every day.

In 1890, an engineer, M. Ridel, devised the wooden box that was used to carry the cheese and helped to send it longer distances, in particular to America, where it became very popular. These boxes are still used today. The variety named Camembert de Normandie was granted a protected designation of origin in 1992.

MARTHA MATILDA HARPER

KEEPING A HEALTHY HEAD OF HAIR

INVENTION: THE RECLINING SHAMPOO CHAIR

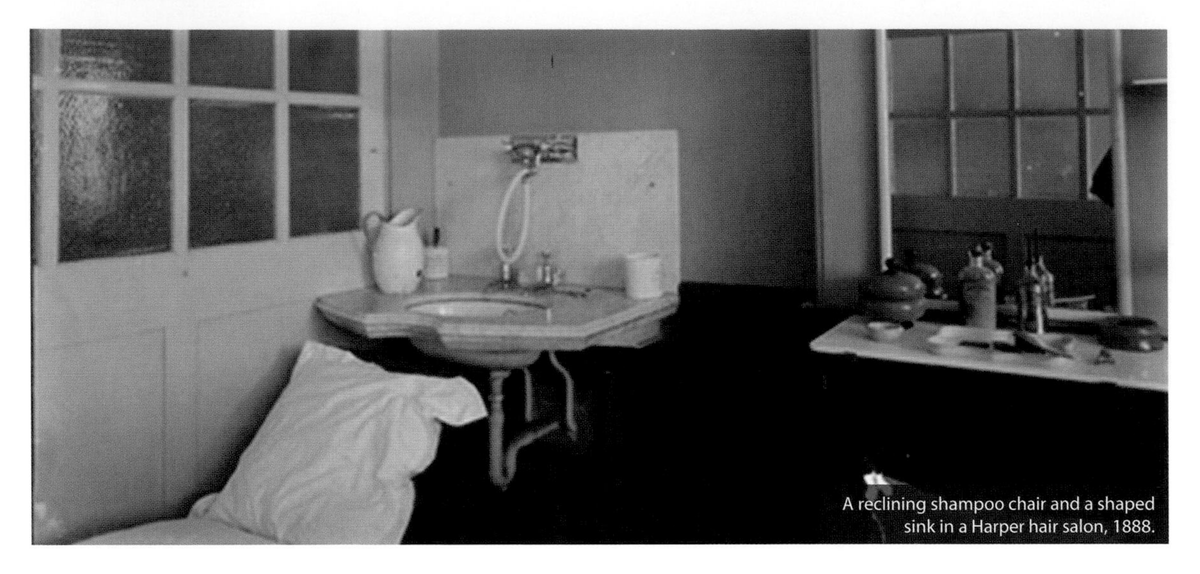

A reclining shampoo chair and a shaped sink in a Harper hair salon, 1888.

Martha Matilda Harper (1857 – 1950) started from nothing. Born in Ontario, Canada, she was just 7 years old when her father sent her away to become a domestic servant for relatives in Orono about 55 miles (88 km) from Toronto. For twenty-two years she worked as a servant in Canada before moving in 1886 to Rochester, New York, to take up another domestic post. Her last Canadian employer had given her an interest in hair, passing on to her all his knowledge on the subject and when he died, he bequeathed her a formula for a hair tonic that he had created.

THE HARPER HAIR PARLOUR

In America, she began the development of a hair tonic of her own, believing that existing hair tonics actually damaged hair instead of keeping it healthy. She had a vested interest in healthy hair as hers stretched all the way to the floor. It would become an important part of the marketing and advertising campaigns she ran later.

Just three years after emigrating to the United States, Martha used her life savings of $360 to open the first public hair salon in the area so that she could market her tonic, calling it The Harper Hair Parlour. At this time, women generally had their hair done at home by their servants. There were independent hairdressers, too, but nothing like Martha Harper's salon.

WINDOW HAIR DISPLAY

The challenge was how to convince women that they should visit her salon, which meant that a good location was absolutely vital. With the help of the people she was working for she secured space in the centrally located Powers Building in downtown Rochester, which was ideal.

In the window she displayed a large photograph of herself and her floor-length hair. The salon took off and women clamored to try out the new service which became known as the Harper Method.

THE RECLINING SHAMPOO CHAIR

The reclining chair was an innovation Martha designed. It meant that her customers could have their hair washed without the discomfort of getting shampoo in their eyes. She also had a sink shaped with a half circle cut out on which the customer could rest her head. Sadly, she failed to patent these. Her customer service was second-to-none and once a woman had experienced it, she was happy to return again and again. She was attracting influential and well-known society ladies and they were happy to spread the word among their friends.

INFLUENTIAL CLIENTS

Among her clients was Susan B. Anthony, the social reformer and women's rights activist who played a leading role in the women's suffrage movement.

The inventor of the telephone, Alexander Graham Bell's wife Mabel was also a customer, as was Grace Coolidge, the wife of the future President of the United States, Calvin Coolidge. It became a place to meet friends and network.

The socialite and social activist Bertha Honoré Palmer was a huge fan of the salon and insisted that Martha should open another in Chicago in time for the 1893 World's Fair, going as far as to promise patronage from twenty-five of her friends.

FRANCHISING THE SALON

The question was, however, how to expand her business with the limited resources she had. She found a solution in the Christian Science Church which she had joined several years earlier after falling ill and being nursed back to health by a Christian Science healer. The church was led by its founder Mary Baker Eddy, and it operated using a network of satellite churches across the country.

The photo Martha Matilda Harper displayed in her shop windows showing herself with her floor-length hair.

Martha decided to adapt a network of satellite salons as a business model to help women in a similar situation to herself to better themselves and to gain financial independence. Poor women, she insisted, would launch the first 100 salons based on hers. It was a very early example of franchising, sixty years before McDonald's founder Ray Kroc introduced it.

The first two franchises opened in 1891 in Buffalo and Detroit, the deal being that the franchisees had to buy her chair and sink and the products they used would be hers. She often loaned these franchisees, known as "Harperites," the money to open because they invariably did not have the necessary funds.

FROM STRENGTH TO STRENGTH

The Chicago salon opened in time for the World's Fair two years later and the business continued to mushroom. Eventually, there were 500 shops across the United States and around the world. Martha also launched a network of training establishments where women were taught the Harper Method and opened a factory in Rochester to manufacture her products.

Clients at the shops included the British royal family, the German Kaiser, and Rose and Joseph Kennedy, parents of the future President Kennedy. While he was involved in the negotiations for the Treaty of Versailles, US President Woodrow Wilson is reported to have traveled every night to the Harper salon in Paris to have a scalp massage. Her hair treatments for men were just another of the innovations she introduced such as the provision of childcare and opening in the evening to accommodate women's busy schedules.

THE HARPER LEGACY

Martha Harper always insisted she would not marry until she was fully in control of the massive empire she had created and finally, at the age of 63, she felt she was sufficiently in control to marry 39-year-old Robert MacBain. He took over the business when she finally stepped aside in 1932 at the age of 75.

Martha died in 1950 at the age of 92, but by that time MacBain had transformed the ethos of her company, introducing chemical dyes and permanents, banned while she was in charge, in place of natural products. He sold the company in the 1970s and the last Harper salon closed in Rochester in 2005.

DIANE HART

LANDMINES AND LINGERIE
INVENTION: CLEARING EXPLOSIVE DEVICES

When the actress Diane Lavinia Hart (1926 – 2002) died at the age of 75, she was remembered for her roles in many of London's West End theaters performing with a flair for sophisticated comedy and wry satire. But, fine actor though she was, she also displayed a talent for litigation. Her biggest payday in the courts was when she won £15,000 in libel damages after a 1981 porn film *Electric Blue 002* illegally incorporated a clip from one of her films, without her knowledge or permission.

However the invention for which she is also known was quite simple but very effective. It was adopted by the War Office during Britain's war with Argentina over the Falkland Islands in 1982. She proposed attaching a farmer's agricultural harrow by cable to a helicopter to clear the live landmines left behind by the Argentine army. The helicopter used the harrow to rake the ground and expose or blow up the mines.

She also tried her hand at commerce. With a friend and fellow actress, Pamela Manson, she set up a ladies' underwear business. Diane was the designer, creating glamorous evening and bridal corsets in combinations of white, gold, black, and silver. She designed the "Beatnix" corselet which was sold on the shelves of Marks and Spencer in the 1960s. She took it to Russia where she reckoned Russian women were starved of good quality lingerie. It sold well.

Diane Hart also dabbled in politics, once attempting to set up a Women's Party. She went so far as to hire Caxton Hall in London for a rally but no more than forty women turned up. She ran in the 1970 general election as an independent candidate in the constituency of South Lewisham but lost her deposit.

Hart is fondly remembered in her later years, weaving on a bicycle up King's Road in London's fashionable Chelsea district, dressed in a full-length mink coat.

A Sea Dragon helicopter tows a minesweeping sled while conducting simulated mine clearing operations.

BEULAH LOUISE HENRY

LADY EDISON

INVENTION: A BOBBIN-FREE SEWING MACHINE

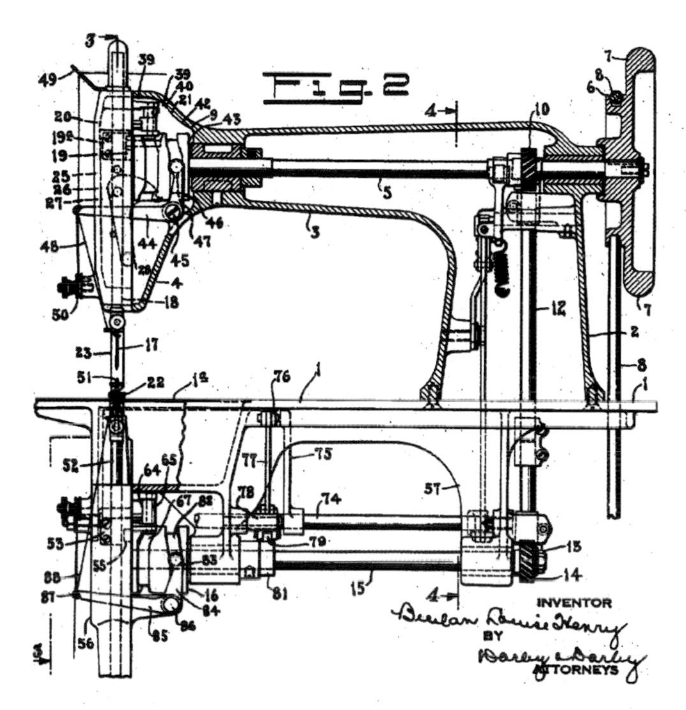

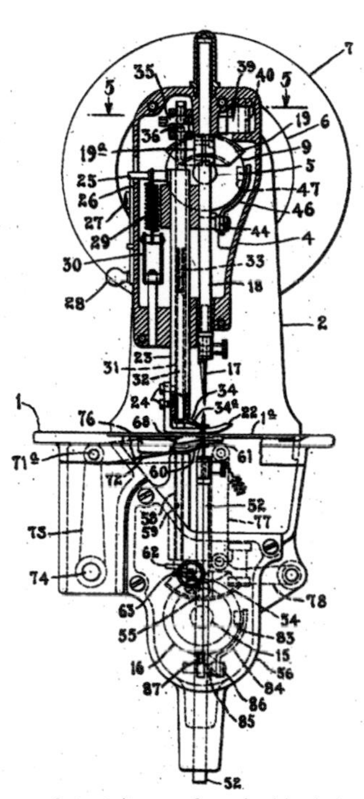

Patent diagram for a double chain stitch sewing machine, Beulah Louise Henry, December 3, 1940.

Legend has it that Beulah Louise Henry (1887 – 1973) began inventing while she was still a child, sketching mechanical gadgets before she was old enough to draw a straight line. She has been granted at least forty-nine patents (the number is the subject of some debate), and is thought to have as many as 110 inventions to her name. She was such a prolific inventor that they called her "Lady Edison."

Born in North Carolina she was granted her first patent in 1912 for a vacuum ice cream freezer. The following year two more were granted for a handbag and a parasol that each had detachable fabric covers in a range of colors so they could be changed to match a particular outfit.

She moved to New York where she established the Henry Umbrella and Parasol Company of which she was president. In 1929 she founded another business, the B.L. Henry Company of New York. Patents continued to be granted to her, one of them for a spring-limbed doll. She found a manufacturer for her parasol invention and they sold 40,000 in two months.

In the 1930s and 1940s Beulah was issued with twelve different patents for typewriter enhancements such as the Protograph, a device that made four copies of a document without the use of carbon paper. She also patented numerous sewing machine improvements, and worked as an inventor for the Nicholas Machine Works in New York. By 1930, she had already sold around forty ideas to big companies.

Beulah never married and lived for most of her adult life in hotels. With little or no technical training, she turned her mind to inventions in a wide range of different areas. "All one needs for inventing," she said, "is time, space, and freedom." Beulah described that when inspiration struck her it came in the form of a fully developed three-dimensional image, "wheels turning and everything." She called it her "inner vision."

Beulah Henry is no longer the most prolific woman inventor, but her tally of patents has only recently been surpassed. "Lady Edison," who passed away in 1973, was arguably the greatest woman inventor of them all.

ERNA SCHNEIDER HOOVER

REVOLUTIONIZING COMMUNICATION
INVENTION: COMPUTERIZED TELEPHONE SWITCHING

Erna Schneider's career trajectory is quite extraordinary. Born in 1926 in Irvington, New Jersey, she studied classical and medieval philosophy and history at the prestigious Wellesley College, graduating with honors and going on to earn a PhD from Yale in philosophy and foundations of mathematics in 1951.

She was professor of logic and philosophy at Swarthmore College between 1951 and 1954 but was unable to make any further progress in her academic career, most likely because she was an unmarried woman. That was remedied, however, when in 1953 she married Charles Hoover Jr. who was employed by Bell Telephone Laboratories in New Jersey. She decided to leave academia and also got a job at Bell.

She was one of the few women in the company and the first to be appointed as technical supervisor. It was a time when switching systems (the systems used to connect telephone calls) were transferring from hardwired and mechanical technology to computer-based technologies. There were problems when a call center was inundated with thousands of calls in a very short time span, overwhelming the unreliable electronic relays and causing the entire system to stop working.

It is said that her revolutionary solution to this problem came to her while she was in hospital for the birth of one of her three daughters. Using her knowledge of symbolic logic and feedback theory, she employed a computer that monitored the frequency of incoming calls at different times. The computer adjusted the call acceptance rate accordingly. Doing this, she eliminated the potential for an overload of processing incoming calls.

In 1971, she was granted one of the first patents for a piece of software and was promoted. The basic principles of her system are still in use today, dealing with the increasing amounts of incoming communications traffic.

In the early days of telephony, companies used manual telephone switchboards, connecting calls with plugs. Dropped calls and busy signals were an annoyance that most could do without. But there would have been a lot more of them if it were not for Dr. Erna Hoover. While working at Bell Laboratories, she invented a telephony switching computer program that kept phones functioning under stressful loads.

GRACE HOPPER

AMAZING GRACE
INVENTION: COMPUTER PROGRAMMING

Lieutenant Grace Hopper operating an early computer with a paper punch tape feed, 1944.

Grace Hopper was born into a fairly well-to-do family in New York in 1906 and was always a bit different to the other girls. While they were playing with dolls, she was indulging her passion for finding out exactly how things worked, taking apart and trying to reassemble all of the family's alarm clocks, for example.

She went to Vassar College in New York, majoring in mathematics and physics and was a brilliant student, often tutoring other students who were having difficulties with the work. Having graduated from Vassar, she began post-graduate studies at Yale from which she earned both her master's degree and a doctorate in mathematics by 1934. Her first job after that was teaching mathematics back at Vassar.

ENLISTING IN THE NAVY

Her involvement with the US Navy began in 1942 when Congress authorized the Navy Women's Reserve as well as a unit called WAVES which stood for Women Accepted for Voluntary Emergency Service. By this time, Hopper was 35 years old, too old to enlist. She was also too slight. At 105 pounds she was 16 pounds too light for her five feet six inches frame.

The other problem was that she was in a protected profession; as a teacher of mathematics she was considered too valuable to go to war. Her argument was that she would be able to contribute more to the war effort in the WAVES than in a classroom and anyway, she insisted, she had always been skinny.

Eventually, after arguing for a year, she managed to persuade the Navy to abandon their normal rules. She obtained a leave of absence from Vassar and enlisted. By this time, she was 37 years old but, naturally, in June 1944 she graduated top of her class of 800 officer cadets.

A BUG IN THE PROGRAM

She began work at the Bureau of Ordinance Computation Project at Harvard under the supervision of the American computer pioneer, Howard Aiken. In his laboratory, design, testing, modification, and analysis of weapons were undertaken. He had a staff of women computors who made calculations using desk calculators.

When Hopper arrived, Aiken is reported to have pointed at a computer he had designed named Mark I and barked at her: "That's a computing engine. I would be delighted to have the coefficients for the interpolation of the arc tangent by next Thursday." It was a daunting task, especially as she was, after all, a mathematician and not a computer programmer.

Hopper had to learn how to program a computer and make a program work, but she approached the problem with her customary determination and eagerness to learn and soon she knew all there was to know about Mark I. While working on it, she discovered that there was a malfunction due to a moth being caught in a relay wire. She removed it and taped it to her report. Since that day, any problem with a program is referred to as a "bug."

Grace Hopper was a pioneer in computer technology and devised the computer language COBOL.

INVENTING COBOL

When the war ended, WAVES were no longer required and Hopper requested a transfer into the regular Navy but her age (she was now 38) worked against her. She continued, however, to serve in the Navy Reserve. In 1949, she joined the Eckert-Mauchly Computer Corporation as a senior mathematician, working in the team that was developing UNIVAC I, the first commercial computer produced in the United States. It was also the first American computer designed for business use.

When Hopper suggested that it would be helpful if a new programming language could be developed that used English, she was shouted down, being told in no uncertain terms that computers are unable to understand English. She refused to let it lie, however, and a few years later published a paper on the subject of compilers, compilers being computer programs that transform source code written in a programming language into another language.

By 1952, she had a compiler that did what she wanted and was operational but nobody really believed her. She was firmly of the belief that data processors ought to be able to write their programs in English, with the computers then translating them into machine code. This marked the start of her invention of COBOL (Common Business Operating Language), a computer language for data processors, which became one of the universally accepted coding languages of the time.

AUTOMATIC DATA PROCESSING

Having retired from the Naval Reserve in 1966 as a commander, she was recalled to active duty in 1967 for a six-month assignment as the Special Assistant of the Navy for Automatic Data Processing. Nineteen years later, she was still there.

She was promoted to captain and Special Advisor to Commander, Naval Data Automation Command in 1973, commodore in 1983 and rear admiral in 1985. She at last retired, very much against her will, in 1986 at the age of 79.

They called her "Amazing Grace," a feisty, eccentric, and very brilliant woman who was in the US Navy for 43 years. This computing pioneer died in 1992 and four years later the destroyer USS *Hopper* was named in her honor.

HARRIET HOSMER

DEFYING ATTITUDES TO WOMEN
INVENTION: TURNING LIMESTONE TO MARBLE

Harriet Goodhue Hosmer (1830 – 1908) was born in Watertown, Massachusetts, and was her parents' only child to survive into adulthood. At school, she began to show an early aptitude for mechanics but she was also encouraged to draw and sculpt and studied anatomy with her father who was a doctor. She also conducted experiments with machines and household devices, although we have no details of what purpose these devices served.

She became fascinated with art, and decided that she wanted to be a sculptor, so she traveled to Rome to study. There, she became involved with a colony of famous artists and writers including Nathaniel Hawthorne and English poets, Robert Browning and his wife Elizabeth Barrett Browning.

Working in the neoclassical style, Hosmer became the most famous and successful American woman sculptor of her day. It was quite extraordinary for a nineteenth-century woman to have a profession such as hers as she said in a testimonial to her own independence:

> *I honor every woman who has strength enough to step outside the beaten path when she feels that her walk lies in another; strength enough to stand up and be laughed at, if necessary.*

In 1879 she was granted a US patent for a process of making artificial marble from limestone. Described by the *New York Evening Post* as the "perfect marble to all appearances as ever was quarried," it was a great commercial success. Harriet Hosmer died in Massachusetts in 1908, having spent her life defying the restrictions that nineteenth-century attitudes placed on women.

Sculptor Harriet Hosmer on top of a step ladder working on a giant statue.

VALERIE HUNTER-GORDON

DIAPER-MAKERS DO BATTLE

INVENTION: THE DISPOSABLE DIAPER

There is a debate as to who actually designed the world's first disposable diaper, the American Marion Donovan or the Scottish woman, Valerie Hunter-Gordon. If it was a matter of the earliest commercially successful disposable diaper, then there would be no argument. Donovan's version evolved into the phenomenally successful Pampers range, still on sale today. Hunter-Gordon's version, the PADDI, disappeared from the shelves in the 1990s.

USING OLD PARACHUTES

While awaiting the birth of her third child, Hunter-Gordon looked forward with dread to the endless chore of washing, drying, and ironing diapers. "I just didn't want to wash them," she later said. "It was awful labor." Realizing that there was apparently no alternative to this drudgery, she resolved to find her own. Initially using old parachute material, she invented a two-piece diaper system that consisted of a biodegradable, disposable pad which was made of cellulose wadding covered in cotton wool. This was worn inside a garment—the PADDI—that was adjustable and was fastened with press studs, thus doing away with the need for a safety pin.

Her diaper proved very popular with other young mothers she knew and she produced around 400 PADDIs seated at her Singer sewing machine at home, toward the end using the newly created material PVC. These were sold to her acquaintances for five shillings each. She had taken the idea to manufacturers but, as with Marion Donovan's disposable diaper, they could not foresee the potential.

NOBEL PRIZE CONNECTIONS

She was granted a patent in the United Kingdom in 1949 but this was only for the garment itself and not for the disposable pad inserts. With the big companies still refusing to help her, it was only by luck that her PADDI made it to the shop shelves. Her father, Sir Vincent Ziani de Ferranti, chairman of the electrical and electronics giant Ferranti, happened to mention his daughter's invention to the Nobel Prize-winning organic chemist Sir Robert Robinson. A contract with his family's firm Robinson and Sons of Chesterfield to manufacture her creation followed and in 1950, the British retail chain Boots began selling the PADDI.

At the same time, she was granted a patent for the United States and around the world. It was demonstrated at the 1952 Ideal Home Exhibition and the BBC selected it as one of the six most interesting products at the exhibition. By the end of the year, 750,000 PADDIs had been sold and sales had risen to 6 million by 1960.

PADDIs vs PAMPERS

The success of PADDIs was only reduced when Pampers launched an all-in-one disposable diaper. In that design, the plastic was disposed of along with the wadding which was environmentally damaging and which had been the object of much criticism by environmentalists.

Valerie Hunter-Gordon was able to escape even more drudgery in the years to come, because she had three more children. Nevertheless, she continued to develop the PADDI and she designed other products, including the innovative Nikini sanitary towel system. This wonderfully inventive and pragmatic lady died in 2016, at the grand old age of 94.

A 1950's advertisement for the PADDI and PADDI-pads system.

IDA HENRIETTA HYDE

SEDUCED BY SCIENCE

INVENTION: AN INTRACELLULAR MICROPIPETTE ELECTRODE

Ida Henrietta Hyde (1857 – 1945) was an American physiologist whose love of biology came from reading the work of the great nineteenth-century Prussian scientist and explorer Alexander von Humboldt. At the time, she was working in a store as a sales assistant, but his writing encouraged her to pursue her education in scientific subjects.

Ida was born in 1857 to German immigrants, Meyer and Babette Heidenheimer who adopted the name "Hyde" when they arrived in America. Their new life had started well, Ida's mother had been running a successful shop. But the shop was destroyed in the 1871 Chicago Fire, and by then Meyer Hyde had left his wife and family. With her husband and store both gone, the family had no means of income.

Ida was forced to leave school at age 16 to become apprentice to a milliner. While working six days a week, she continued her studies with evening classes at the Chicago Athenaeum. Sadly, however, her mother wanted her son to receive a good education too, and in line with the times, she cared rather less about the education of her three daughters.

BEGUILED BY BIOLOGY

When Ida reached the age of 24, with the words of Alexander van Humboldt ringing in her ears, she took a year off work and sunk all her savings into a year of college education. After that, she passed the county teachers' exam and three years later the Chicago teachers' exam. It was during a summer school at Martha's Vineyard, still with a fascination for biology, that she decided to become a scientist.

In 1889, she enrolled at Cornell University and two years later graduated with the intention of studying medicine. She changed her mind, however, and instead accepted a graduate fellowship at the women's college, Bryn Mawr, Pennsylvania, where she worked alongside eminent scientists such as the American evolutionary biologist Thomas Hunt Morgan, who won the Nobel Prize in 1933 and Jacques Loeb, the German-born American biologist.

The Association of Collegiate Alumnae (ACA) in Washington DC was an organization set up in 1882 to further the goal of improving women's educational opportunities. In 1893, partly because of her research on the nervous system of jellyfish at Woods Hole Biological Laboratory, Massachusetts, Ida received a fellowship from the ACA and was invited to study in Europe at Strasbourg University. (Strasbourg at the time was part of Germany.)

FIGHTING PREJUDICE AND BIAS

At Strasbourg, she became the first woman in German academic history to try to matriculate for an advanced degree in natural science or mathematics. To enrol, she had to petition the German government and obtain the permission of the faculty but too many people opposed a woman undertaking such studies, and she was forced to withdraw her application.

Ida Henrietta Hyde in her laboratory at the University of Heidelberg, 1896.

Instead, she took a PhD at Heidelberg University, only the third women to obtain a doctorate from that institution. But even there, she was forbidden to attend lectures or practical sessions and had to rely on notes prepared for her by laboratory assistants. She constantly faced discrimination on gender grounds and continued to fight prejudice and bias against women throughout her career.

In 1899, she was employed by the University of Kansas in Lawrence to establish a physiology department. It was very unusual at the time for a woman to be made a professor at a co-educational establishment. In the summer, she studied at Rush Medical School in Chicago, pursuing her MD which she finally received in 1911.

THE MICROPIPETTE ELECTRODE

Ida Hyde researched and taught but she was also an inventor and an innovator, developing instruments and equipment throughout her career. Her most well-known invention was the intracellular micropipette electrode in 1921. In her research, she had discovered that electrolytes in high concentrations affect cell division which helped her examine the minute differences in electrical potential within cells.

The intracellular micropipette electrode could be used to stimulate cells at the micro level while recording electrical activity within the cell without disturbing the cellular wall. The device was a revolutionary invention in neurophysiology, but she was never fully credited with being its inventor. A similar microelectrode invented twenty years after hers won a Nobel Prize nomination for its inventor in the 1950s.

Ida left the University of Kansas in 1918 and never returned, retiring to California but continued with her research. She died of a cerebral hemorrhage in 1945, at age 88, and was buried close to the laboratory at Woods Hole.

THE GREAT CHICAGO FIRE

In 1871, the Great Chicago Fire burned from Sunday, October 8, until Tuesday, October 10. It began around nine on the Sunday evening at a small barn that was owned by a family named O'Leary, although the exact cause has never been determined. One theory suggests that a cow may have knocked over a lantern and another suggests that the lantern was knocked over by a group of men who were gambling in the barn.

It spread quickly, helped by the wood of which most of the city's buildings were constructed. Most buildings also had highly flammable tar or shingle roofs and even the sidewalks and many roads used wood in their construction. Drought conditions and southwest winds that fanned the flames and carried sparks into the center of the city further compounded the situation. Worse still, the city's small fire-fighting force was sent to the wrong location.

It was hoped initially that the Chicago River would provide a firebreak but the buildings and warehouses lining it were of wood and quickly caught fire. The flames jumped the river and set alight the gas works on the other side. The city's waterworks burned down, leading to the city's water mains running dry. The fire raged on from block to block unchecked.

It finally began to rain on the evening of October 9 and the fire burned itself out the following day. By then, it had destroyed 3.3 square miles of the city and is believed to have killed up to 300 people, although the true number of casualties has never been known. More than 100,000 Chicago residents were left homeless.

The Great Chicago Fire aftermath, 1871.

HYPATIA OF ALEXANDRIA

OUT-THINKING THE PHILOSOPHERS
INVENTION: THE ASTROLABE AND THE HYDROMETER

The extraordinary Hypatia of Alexandria was a Greek astronomer and philosopher who lived in fourth century Egypt. She is credited with being the first woman mathematician and some have claimed that she invented the astrolabe, the essential navigational instrument used by sailors until the invention of the sextant. The astrolabe enabled the user to identify stars and planets, to work out local latitude, to survey, and to triangulate.

The claim that Hypatia was the inventor of this invaluable tool comes from letters written by Synesius, one of her students, in which he also claims that Hypatia invented the hydrometer, an instrument that measures the specific gravity of liquids:

> *It has notches in a perpendicular line, by means of which we are able to test the weight of the waters … Whenever you place the tube in water, it remains erect. You can then count the notches at your ease …*

Born around 350 AD, Hypatia was the only daughter of the mathematician, Theon of Alexandria. She was educated in Athens, and around the year 400 AD was appointed head of Alexandria's Neo-Platonic School teaching students the philosophy of Plato and Aristotle. In the seventh century, she was described by an Egyptian Coptic bishop, John of Nikiû, as a pagan who was "devoted at all times to magic, astrolabes, and instruments of music and she beguiled many people through her satanic wiles." Others saw her however as an exemplar of all that was good, as described by her contemporary, the Christian historian, Socrates of Constantinople:

> *There was a woman at Alexandria named Hypatia, daughter of the philosopher Theon, who made such attainments in literature and science, as to far surpass all the philosophers of her own time.*

Sadly, Hypatia came to a rather sticky end in the year 415 AD. She was a counselor to the Roman governor Orestes, and people blamed her for his refusal to settle a feud with the Bishop of Alexandria. She was carried away by an angry mob who, according to fifth-century writer Socrates Scholasticus, "murdered her with tiles," dragging her through the streets until she was dead. A gruesome death for a great woman.

An astrolabe is an elaborate instrument used by astronomers and navigators to identify stars and planets.

NANCY JOHNSON

HAND-CRANKING A FROZEN DELIGHT
INVENTION: THE ICE CREAM MAKER

For most of the nineteenth century, ice cream was a pleasure enjoyed during a hot summer mainly by the wealthy upper classes. That changed in America in 1843 with Nancy Johnson's invention of a hand-cranked ice cream making machine. Johnson's machine consisted of an outer wooden pail filled with crushed ice.

The ice cream mixture to be frozen was contained in an inner cylinder made of tin or pewter. A lid was bolted on to the contraption, the cranking handle was inserted into the top of the lid and turned. As the ice cream mixture came into contact with the side of its container, it gradually started to freeze. The cylinder for the ice cream mixture could also be split into two separate sections so that two different flavors of ice cream could be produced at the same time.

Sadly, Johnson did not make a fortune from her machine. Unable to afford to patent it, she sold the right to do so for just $200 to a man named William Young who marketed the machine as the Johnson Patent Ice Cream Freezer. For many years, anyone who wanted to make ice cream would turn to Nancy Johnson's invention, earning Young a small fortune. The basic principle that she developed is still in evidence in the ice cream makers that are sold today.

Hand cranking the handle of an old fashioned ice cream maker.

THE HISTORY OF ICE CREAM

The Italian Duchess Catherine de' Medici married the future Henry II of France in 1533, bringing with her to France Italian chefs who brought with them recipes for flavored ices and sorbets. But, the first recipe for flavored ices first appeared in French in a 1674 book by Nicholas Lemery and there were recipes for sorbets in a 1694 edition of Antonio Latini's *Lo Scalco alla Moderna* (*The Modern Steward*).

Other publications described how to make flavored ice. In England, ice cream recipes began to appear in the eighteenth century. It was mentioned in the 1744 edition of the *Oxford English Dictionary* and a recipe appeared in cookery writer Hanna Glass's *The Art of Cookery Made Plain and Easy* in 1751.

In America, Quaker colonists arrived with ice cream recipes and during the colonial period it was on sale in confectionery shops in New York and other cities. George Washington is said to have spent $200 on ice cream in the summer of 1790. Ice cream freezers, such as the one invented by Nancy Johnson helped make it still more popular.

In England, ice cream was still something of an expensive item, reserved for those with access to an ice house or a freezing machine, until the Swiss émigré Carlo Gatti set up a stand outside London's Charing Cross railway station in 1851, selling scoops for a penny. Gatti stored his ice in an "ice well" at Regent's Park but as his business grew, he began importing ice from Norway.

Agnes Marshall introduced the edible ice-cream cone in 1888, calling it a cornet and it became even more popular when it was sold at the 1904 World's Fair in Missouri. There was no stopping ice cream now and by the second half of the twentieth century it was a popular treat all round the world.

REBECCA JOHNSON

IOWA'S POULTRY QUEEN
INVENTION: AN EGG-HATCHING INCUBATOR

An antique illustration from 1900 of an egg-hatching incubator.

In the late nineteenth and early twentieth centuries, Rebecca Johnson (1853 – 1921) became known as Iowa's Poultry Queen, but her involvement with chickens only came out of necessity. Born Rebecca Salena Veneman in Illinois, it was when her first husband died, leaving her with three young children, that she considered trying to support her family with poultry farming.

Knowing next to nothing about chickens, she had to learn on the job, building a warm house to keep her hens in cold weather and feeding them vegetables so that they would lay all winter. Before long, she was making more than enough to feed her family and buy feed for her cattle and pigs as well. She remarried and had another four children, increasing the pressure on her finances, but when her second husband fell ill, she decided it was time to become serious about poultry.

The first incubator she invented was designed after she watched a hen for many hours. She noted that the hen stood up every now and then and moved her eggs around. Johnson made marks on the eggs with a pencil to confirm that the hen was actually turning them and so this became a part of the way her incubator worked.

Furthermore, the incubator had a thermostat that responded to changes in temperature. In the early versions, when the temperature changed, an alarm bell rang but in later versions, the thermostat actually raised or lowered the temperature. Her incubator alarm was patented in 1908. With this first incubator she could hatch around 5,000 chicks in one season. By 1909, she was earning $300 a month and her incubator could hatch 2,500 chicks a day.

AMANDA T. JONES

PSYCHIC PHENOMENA AND FOOD PRESERVATION

INVENTION: A VACUUM METHOD OF CANNING

Born in East Bloomfield, New York, Amanda Theodosia Jones (1835 – 1914) began teaching at the age of 15 in a one-room schoolhouse while continuing with her education at a local high school. In 1854 at age 19, Amanda became a writer and enjoyed some success with her early poems and songs being published in *Frank Leslie's Illustrated Newspaper*.

She was fascinated by spirits and psychic phenomena, claiming that spirits talked to her and that she could predict the future. In 1859 she was diagnosed with tuberculosis from which it took eighteen months to recover. She moved to the countryside, writing as well as communing with the spirit world, and trying to regain her health. In 1869, when she was 34, a dream persuaded her to make changes in her life.

In the nineteenth century canning had had a long problematic history in which fresh fruit and vegetables were rendered mushy and tasteless by the process which at the time involved cooking the food. Amanda dreamed up her method of food canning while dozing off during an air bath treatment.

In 1872 she enlisted the help of Professor Leroy C. Cooley to help develop her idea and in the following year she was granted seven patents for the process, two of which listed her as sole inventor. In her process, sealed, filled jars were steamed until the heat forced the contents of the jars to expand, extracting the air, and creating a vacuum inside. The process that she invented made it possible to can uncooked fruit and vegetables all year round.

To get her jars of fruit and vegetables to market, she launched the Women's Canning and Preserving Company in Chicago. Amanda did everything from training to sales and even sold stock in her business. After the first year, she made a profit and things were going well. However, the investors gradually took over the business, and Amanda was forced out.

But Amanda continued inventing, being awarded further canning patents in 1903, 1905, and 1906. She also began investigating oil burners, gaining patents in that field in 1904, 1912, and 1914. Her memoir, *A Psychic Autobiography*, was published in 1910 and as the title suggests focused on her psychic experiences. She died in 1914 in Brooklyn, New York, at age 78.

Patent diagram 794940 by Amanda T. Jones, July 18, 1905, for an apparatus for preparing food products.

Jars of preserved fruit and vegetables.

ELDORADO JONES

THE IRON WOMAN
INVENTION: AN AIRPLANE MUFFLER

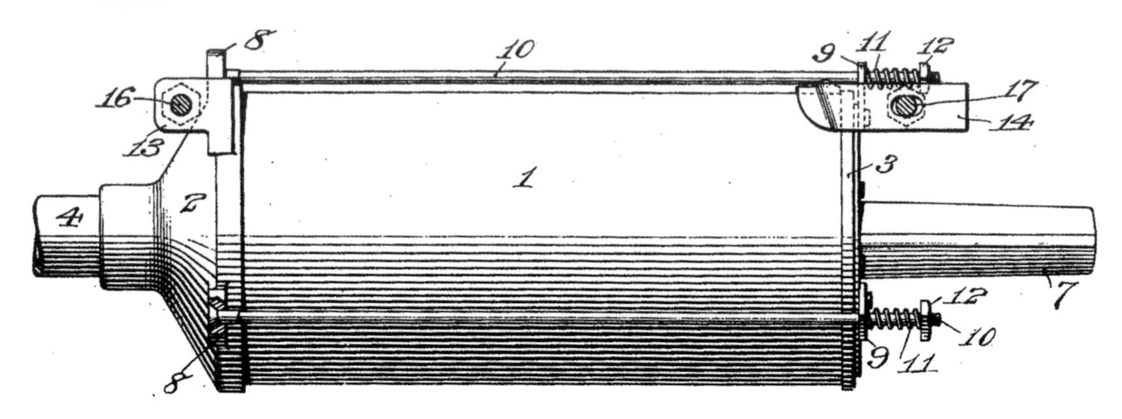

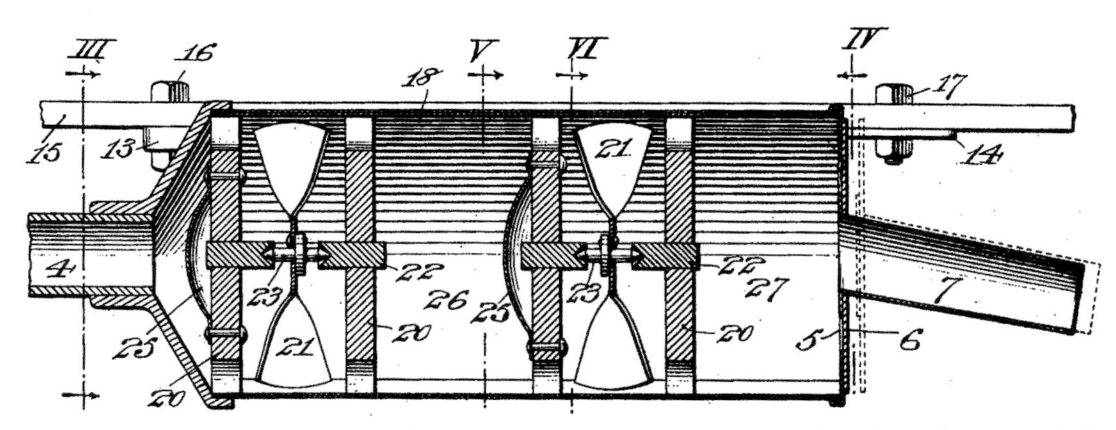

Patent diagram for an improved exhaust muffler, Eldorado Jones, July 31, 1919.

Born in Palmyra, Missouri, Eldorado Jones (1860 – 1932) came to invention late in life. She had been a teacher after her father Alonzo Jones deserted the family when they moved to St. Louis, Missouri, his vanishing act perhaps explaining why she never had much time in her life for men. Neither did she have much time for teaching. She hated it and soon decamped to Chicago where she found work as a stenographer for an insurance company.

She began to invent, and in 1913 opened a factory in Moline, Illinois, that employed only women over 40 years of age and was totally dedicated to manufacturing her inventions. Soon, the factory was turning them out—small, lightweight irons, a traveling ironing board that had a compartment in which a flat iron could be stored, and a collapsible hat rack. Men were not permitted anywhere near her business and when some tried to buy it from her she took great pleasure in rejecting their offer, and began to earn her reputation as the Iron Woman.

In 1919 she invented an airplane muffler, and applied for a patent which was finally granted in 1923. Her muffler was similar in concept to a car muffler and was designed to diminish the noise made by the exhaust of an airplane engine, without reducing power. She had to find financial backing which led her to New York but no one wanted to become involved.

Whether this might have been because of her attitude toward men is unknown but her money was rapidly running out. In fact, she was so short of cash that she applied for welfare aid. By 1932, not only had her money run out but her energy had, too. A neighbor found the Iron Woman dead in bed, at age 72.

MARJORIE STEWART JOYNER

MAKING WAVES

INVENTION: A PERMANENT WAVE MACHINE

In the 1920s, African American women who wanted to straighten their hair could only use a curling iron that had been heated on the stove. It took a lot of time and was extremely frustrating business as only one iron could be used at a time. Marjorie Joyner (1896 – 1994) resolved to find a better way. Thinking laterally, she imagined a number of curling irons arranged above a woman's head straightening all of her hair at the same time. She said later:

A beauty school student attaches the curlers from a permanent wave machine to a woman, 1945.

> *It all came to me in the kitchen when I was making a pot roast one day, looking at these long, thin rods that held the pot roast together and heated it up from the inside I figured you could use them like hair rollers, then heat them up to cook a permanent curl into the hair.*

It was an elegant solution to an age-old problem and her device not only straightened hair, it could also create a permanent wave. She is often claimed to have been the first African American woman to be granted a patent.

Born in Monterey, Virginia, Marjorie Joyner was the granddaughter of a slave and a slave-owner. In 1912, interested in cosmetology, she moved to Chicago to study the subject, graduating in 1916 from the A.B. Molar Beauty School, the first African American woman to graduate from the institution. It led her to work with Madam C.J. Walker, the African American hair and beauty entrepreneur. Walker employed Joyner as national advisor, managing 200 beautician schools across the country.

She was responsible for ensuring that Walker's hair stylists did their door-to-door selling job properly. It has been estimated that Marjorie Joyner taught around 15,000 stylists in a career that lasted fifty years. Unfortunately, however, Madam Walker's company took the patent for Joyner's machine and also took all the money. Joyner received next to nothing.

Joyner helped to draft the first laws on cosmetology for the state of Illinois and her friendship with First Lady Eleanor Roosevelt led to her involvement in the founding of the National Council of Negro Women and various other organizations and agencies to advance the cause of black women. Marjorie Joyner died at the age of 98.

ANNA KEICHLINE

DEFYING GENDER POLITICS
INVENTION: THE "K BRICK"

The building block (1927), better known as the "K Brick," a forerunner of today's concrete block.

Born in Bellefonte, Pennsylvania, in 1889, Anna Wagner Keichline—architect, inventor, and World War I Special Agent—was a woman who did not let the gender politics of the era get in her way. Anna's parents encouraged her interests in carpentry and mechanics, not normal pastimes for a girl in the early twentieth century, and bought her a home workshop fully equipped with a fine set of carpentry tools.

By the time she was in her early teens, her carpentry skills were well known and she even won a local prize for an oak table that she designed and built. Following her graduation from high school in 1906, she went to Pennsylvania State College to study mechanical engineering. The following year she enrolled at Cornell University to study for a degree in architecture, an ideal occupation she reasoned, as women knew how a house worked better than men.

In 1920, she became the first registered woman architect in Pennsylvania and designed buildings there as well as in Washington DC and Ohio. She also began applying for patents. Her first, for a space-saving sink and washtub, had been granted in 1912 and in 1924, she obtained one for a kitchen design that had sloping work surfaces and glass fronted cabinets. A 1929 patent she was granted was for a folding bed in a small apartment to provide more space during the day.

Of her seven patents, the best known was for a building component she called the "K Brick." It was a cheap, light, fireproof clay brick designed for use in hollow wall construction. Its cavities could be filled with insulating or sound-proofing material and it was a precursor of the modern concrete block. It was patented in 1927 and she was honored for it in 1931 by the American Ceramic Society.

However, her architectural designs remain the centerpiece of her fascinating accomplishments. In Bellefonte her designs include the Plaza Theater, the Cadillac Garage and Apartments, the Harvey Apartments, and several private homes.

One of the lesser known aspects of Anna Keichline's career was her role as a special agent during World War I, working for the Military Intelligence Division in Washington DC. Anna Keichline, who was never intimidated by being a woman in a man's world, died in 1943 at the age of 54.

MARY BEATRICE DAVIDSON KENNER

MANAGING FEMININE HYGIENE
INVENTION: THE SANITARY BELT

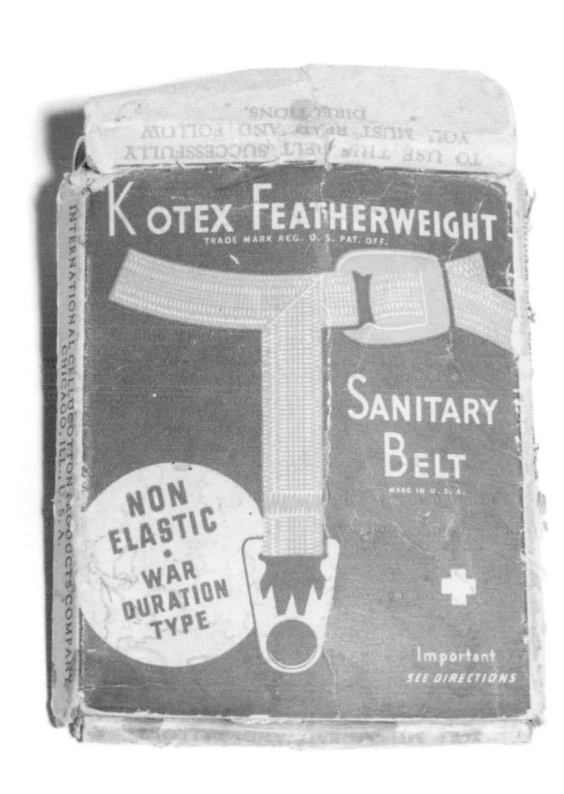

A Kotex sanitary belt, one of several objects from the Museum of Menstruation, New Carrollton, Maryland.

The African American inventor Mary Beatrice Davidson Kenner, who was born in Monroe, North Carolina, in 1912, was a flower arranger by profession but she invented the sanitary belt in 1936, although it was not patented until 1956. She faced not just discrimination because of her sex but also because she was black, companies refusing to work with her to produce her invention, even though it would be of great benefit to women.

Tampons were frowned upon in her time but the alternative was a cloth or rag which was often uncomfortable and which could also be unhygienic. In fact, this method was so inconvenient that women using it usually had to stay home during that particular time of month. Her invention consisted of a belt that stretched around the waist with a section stretching down over a woman's intimate parts and linked to the belt at the back. In this way, the pad was kept in position.

Between 1956 and 1987, she was granted five other patents for household and personal items including a carrier attachment for an invalid walker, a bathroom tissue holder, and a back washer mounted on a shower or bathtub wall.

ELIZABETH KENNY

THE NURSE WHO CONFRONTED CONVENTION
INVENTION: THE SYLVIA STRETCHER

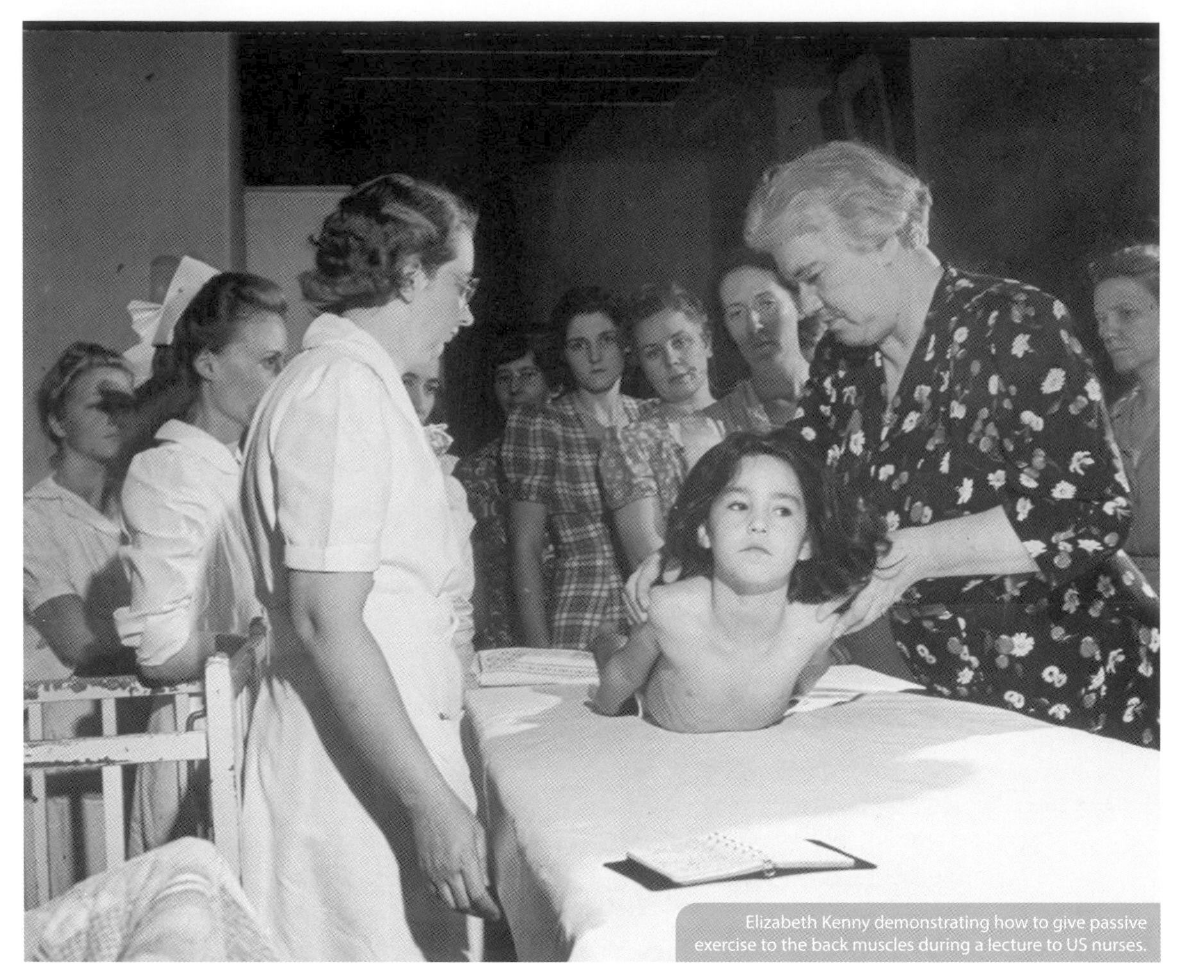

Elizabeth Kenny demonstrating how to give passive exercise to the back muscles during a lecture to US nurses.

Born in New South Wales, Australia, in 1880, Elizabeth Kenny had Irish ancestry, her father was a farmer who had emigrated from Ireland. She became interested in medicine at the age of 17, when she broke her wrist and was taken to the surgery of Dr. Aeneas McDonnell in Toowoomba, Queensland. It was the start of an association with McDonnell that lasted until his death. She worked as an unaccredited nurse until her mid-20s and was trained on the job, never receiving any formal medical training.

ASSIGNED TO DARK SHIPS

Acting as a broker between farmers and markets in Brisbane, as well as nursing, she earned enough money to open a cottage hospital at Clifton, south of Toowoomba. In 1910, she treated her first case of polio. She used an unorthodox and controversial treatment of applying hot compresses to the paralyzed legs as well as weights made from woollen blankets. The patient responded well and several other patients recovered and suffered no serious after-effects.

During World War I she volunteered as a nurse and was assigned to "dark ships," vessels that ran between Australia and England with their lights off, carrying war equipment and soldiers one way and wounded soldiers and trade goods on the return voyage. She made sixteen such return voyages as well as one around the world. Toward the end of the war, she was promoted to "sister," equivalent to first lieutenant in the Australian Army Nurse Corps. It was a title she used for the remainder of her life.

THE SYLVIA STRETCHER

She returned to Queensland after the war and worked in hospitals during the 1919 flu epidemic. She continued her nursing work from her mother's home and it was at this time that she invented Sylvia's stretcher. Sylvia was the daughter of a friend who was run over by a horse-drawn plow. Sister Kenny was summoned and seeing that a stretcher was needed, she improvised one from a cupboard door. The girl was strapped onto the door and carried, with Kenny accompanying her, the 26 miles to the surgery of her friend Dr. McDonnell in Toowoomba.

Elizabeth Kenny, 1950.

It was Kenny's solicitous care during the journey that helped in no small way in Sylvia's complete recovery from such a serious incident in which both her legs had been broken and she had lost a couple of toes. McDonnell suggested that she should make the stretcher available in Australia's more remote regions. She improved it and began marketing it as "Sylvia's Stretcher" in Australia, Europe, and the United States.

FOUNDATIONS OF PHYSIOTHERAPY

The charitable Sister Kenny presented the idea for the stretcher to the Queensland Women's Association so that they could take out the patent on it and supplement their funds from it. But, they insisted that Sister Kenny hang on to the stretcher and the patent and it was a good thing that she did. The royalties she made from Sylvia's Stretcher supported her long and ultimately successful struggle to have her unique polio treatment accepted.

Kenny eventually gained recognition in Australia, establishing several clinics throughout the country. Her principles of muscle rehabilitation went on to become one of the foundations of physiotherapy. In 1921 she even met President Franklin D. Roosevelt who contracted polio but had learned to walk with braces and a cane.

Some have called Sister Kenny the founder of modern physical therapy. While she may not have been quite that, her contribution to the treatment of patients with poliomyelitis was unique. She led many to change their approach to polio patients.

ENDURING LEGACY

Actor Alan Alda credits the Sister Kenny treatment for his complete recovery from polio as a young boy, and actor Martin Sheen regained the use of his legs after contracting polio as a child using Sister Kenny's method. Her life story was told in the 1946 film *Sister Kenny*, portrayed by Rosalind Russell, who was nominated for the Academy Award for Best Actress. But her most enduring legacy is the Minneapolis Sister Kenny Rehabilitation Institute established in 1942 which merged with the Courage Center in 2013 to become the Courage Kenny Rehabilitation Institute.

Suffering from Parkinson's disease, Sister Kenny died on November 30, 1952, after four days in a coma following a cerebral thrombosis.

PHYLLIS KERRIDGE

A WONDERFUL FEMALE BRAIN

INVENTION: THE MINIATURE pH ELECTRODE

In 1925, chemist and physiologist Phyllis Margaret Tookey Kerridge published a paper that discussed her invention of a glass electrode designed to analyze biochemical samples such as blood. She had invented the tool while working toward her PhD at University College London, with a thesis entitled *Use of the Glass Electrode in Biochemistry*. For the research she was doing at the time, she needed a tool sufficiently tiny to fit into narrow layers of tissue.

pH electrodes already existed but the signals they emitted were very weak. Kerridge solved the various issues involved by creating an improved apparatus, overcoming the high resistance of the glass, the danger that such a delicate tool would easily break, and the risk of it short-circuiting. She invented a miniature pH electrode whose heat-treated platinum component gave a much greater signal than previous devices. It gave accurate measurement even when there was not a great deal of fluid present.

Kerridge, who had been born in Bromley in Kent in 1901, had a wonderful technical brain. In the early 1930s, she worked with scientific instrument maker, Robert W. Paul. He was looking for someone to undertake rigorous physiological testing of a non-invasive respirator called the "pulsator," created by Sir William Henry Bragg. Kerridge's work made improvements to the design that helped to reduce its size and its complexity.

In the late 1930s, Kerridge also worked at the Royal Ear Hospital. She was a lover of music and this made her particularly sympathetic to people who were unable to hear. She was heavily involved in establishing hearing aid clinics for the deaf and was given funding in 1936 by the Medical Research Council to test the hearing of London schoolchildren.

Kerridge died in 1940, at the age of just 38.

A silver/silver chloride reference electrode and pH glass electrode.

THE US PATENT ACT OF 1790

The Patent Act of 1790 was the first statute concerning patents passed by the federal government of the United States. It defined the items to which a patent should apply as "any useful art, manufacture, engine, machine, or device or any improvement there on not before known or used." To the applicant was granted by the act the "sole and exclusive right and liberty of making, constructing, using and vending to others to be used" of his invention.

Patent Board members, also known as Commissioners for the Promotion of Useful Arts (the Secretary of State, the Secretary of War, and the Attorney General), were ascribed the authority to grant or refuse a patent after having considered whether the invention or discovery was "sufficiently useful and important."

A patent cost about four or five dollars in total, and the duration of a patent could be any length up to 14 years. The first patent was granted to Samuel Hopkins of Baltimore, Maryland, for his invention for "Making Pot and Pearl Ashes," potash.

MARY DIXON KIES

MAKING HATS TO GET AHEAD

INVENTION: WEAVING STRAW WITH SILK

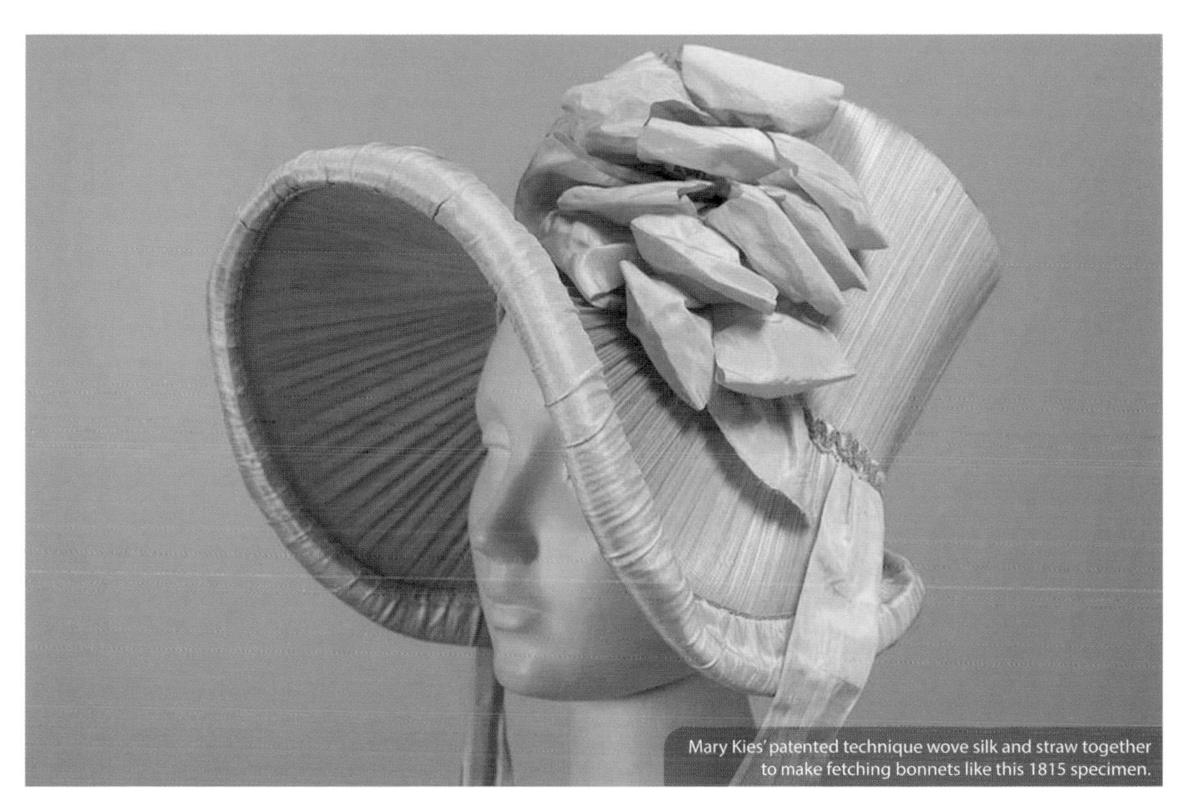

Mary Kies' patented technique wove silk and straw together to make fetching bonnets like this 1815 specimen.

The Napoleonic Wars that lasted from 1803 to 1815 caused huge problems to all the nations involved and trade between some of the richest countries in the world came to a standstill with serious consequences. The United States put an embargo on all trade with France and Great Britain, creating shortages of all kinds of goods. In fact, it was a disaster for the American economy, exports dropping from $108 million in 1807 to a mere $22 million the following year. One solution was for the United States to manufacture some of these imported items itself.

In fashion, France reigned supreme but the embargo prevented the products of French designers making it to the United States. Hats were, at the time, an essential fashion accessory for women and it fell to some American women to come up with new ideas to enable them to produce headgear that was the equal of the French.

The first to innovate in hat-making was a New England woman, Betsy Metcalf who in 1798, after seeing a straw hat in a shop window that she was unable to afford, invented a method of braiding straw that became popular. In effect, she created a whole new industry because not only did she employ girls and women to make hats, women could now make hats at home, using local resources. She could have sought a patent for her invention as the Patent Act had been passed in 1790 but it was complicated for women at the time when they could not even own property. Betsy Metcalf, like many other women inventors, chose not to bother.

Mary Dixon Kies chose otherwise and on May 5, 1809, she was granted the first US patent given to a woman. Her invention was a new technique for weaving straw with silk and thread. Her technique was invaluable in making cost-effective work bonnets. The hat-making industry took off in Massachusetts and in 1810 it is estimated to have been worth $500,000 which in today's terms amounts to more than $4.7 million.

Sadly, Mary Kies never made any money from her invention and died penniless in 1837, at age 85.

MARIE KILLICK

GETTING INTO THE GROOVE
INVENTION: THE TRUNCATED SAPPHIRE STYLUS

Close-up of a record player stylus and arm.

Marie Killick's origins are a little obscure. We know she was born Marie Louise Benson, probably in 1918, and she lived in Belgium and Holland as a young girl where she learned about her father's profession, precision engineering. In the late 1930s, she started a small company in Putney, London, that manufactured sound recording equipment. She developed a specialized system for producing record discs that required a precision engineered stylus, and in 1945 was granted a patent for "Improvements Relating to Styli for Sound Reproduction."

Marie Killick became wealthy from her invention, being awarded a contract in 1946 to produce styli for numerous outlets. Business that amounted to around £1 million in today's money. American orders also began to flood in. The Killicks were living the high life, and Marie even rejected an offer of £750,000 from the record company Decca to buy her business.

But when the banks refused to loan her £2 million to buy more equipment, the company ran into problems as they were unable to supply the massive orders from the United States. The business began to fail, and by October 1949 there were not even enough spare funds to renew the patent for the stylus. Marie needed to get her patent back and in September 1951 she scraped together enough money to renew it.

Suddenly she learned that the electronics company Pye were selling what they called "universal" styli for 78s and the new microgroove records. A direct competitor, Pye were clearly plagiarizing Marie's design. She took legal action in 1954, even selling jewelry, her car, and what remained of her old company to fund the lawyer's fees.

The Pye court case dragged on endlessly and was not settled in her favor until June 1958, by which time Marie had run out of money. Worse still, there were rumors that she would be lucky to get even £5,000 in compensation. In September 1959, she was declared bankrupt, and in 1961 she finally gave up the legal battle.

She and her family vanished from the headlines after August 1961, and Marie Killick died in 1964. She would have been gratified to know that Pye also ran out of money, and only lasted another three years before being bought out by Dutch electrical giant Philips.

ELIZABETH S. KINGSLEY

DECIPHERING THE HIDDEN SOLUTION
INVENTION: THE DOUBLE-CROSTIC PUZZLE

Crosswords exercise our little gray cells and test our vocabularies and general knowledge.

The Wellesley-educated Elizabeth Kingsley was born in Brooklyn and worked as a teacher there. In 1933, she invented what she called the double-crostic crossword puzzle. It is a hybrid somewhere between a cryptogram and a crossword puzzle. It normally consists of two sections, the first a set of lettered clues and the second a series of numbered blanks and spaces. The player answers the clues and fills in the answers according to the numbers allocated and in the order of them. Therefore, if the answer is "GUN" and it is given the numbers 6, 3, and 13, "G" will go in to space 6, "U" will go into space 3, and "N" will go into space 13. This is done with the other clues and answers and at the end the answers will reveal a hidden word, quote, or phrase.

Kingsley spent what must have been six brain-numbing months compiling a hundred of these puzzles. She passed them to the *Saturday Review of Literature* and the editor signed her up to produce one a week, the first appearing on March 31, 1934. She compiled puzzles for the *New York Times* from May 9, 1943, until December 28, 1952, and also published books of her puzzles. The style of puzzle she invented remains popular to this day.

INVENTING THE CROSSWORD PUZZLE

It is hard to say when the first crossword puzzle appeared. Crossword-like puzzles are said to have surfaced in a short-lived publication, *The Stockton Bee* that ran from 1793 to 1795. The actual name "cross word puzzle" was first used in 1862 in the illustrated children's periodical, *Our Young Folks*. In 1913, Arthur Wynne, an English journalist who had emigrated to the United States from Liverpool in 1891, published what he called a "word-cross" puzzle in the *New York World*. It was the first example of the puzzle that we know today and Wynne is often said to have been the inventor of the crossword puzzle. He introduced elements such as the use of horizontal and vertical lines to create the boxes and he later introduced black boxes to separate the words. A few weeks later, the name of Wynne's puzzle changed to "Cross-Word" as a result of a typing error and that is the way it has stayed.

MARGARET KNIGHT

WRAPPING UP THE GROCERIES
INVENTION: THE FLAT-BOTTOMED PAPER BAG

Margaret Knight's paper bag machine is on display at the Smithsonian's National Museum of American History.

Margaret Knight (1838 – 1914) was probably the most famous nineteenth-century female inventor. She was born in York, Maine, and lost her father at an early age after which the family moved to Manchester, New Hampshire. She was unlike other little girls, she thought dolls silly, preferring instead to whittle wood and build things.

She came up with her first invention at age 12 in a cotton mill when a broken thread snagged a moving shuttle which flew off the loom and stabbed a girl. Margaret worked out that the injury could have been avoided if the machine had shut down instantly. She made a safety device that accomplished this and it was adopted by a number of mills, although she never patented it.

In 1867 she moved to Springfield, Massachusetts, where she found a job with the Columbia Paper Bag Company. It was here that she invented the flat-bottomed paper bag with which we are all so familiar. Her patent application was not without drama, however. She had built a wooden model of her machine, but for a patent to be granted, she had to have a working iron model. A man named Charles Annan from the machine shop where the iron model was being made, stole her design, applied for a patent, and was granted it.

Margaret filed a patent interference lawsuit which she won, and in 1871 she was finally awarded the patent. She founded the Eastern Paper Bag Co. to manufacture her paper bags. She received royalties but it is unclear whether she actually made much money from her venture.

No one really knows how many inventions Margaret Knight can be credited with, as many of her inventions were never patented. There were probably twenty-two patented inventions, either in her name alone or in partnership with men. Eight of those dealt with rotary engines and the others included a numbering machine, a sole cutting machine, and a window frame with sash.

Margaret Knight died in 1914, at age 76. She was inducted into the National Inventors Hall of Fame in 2006, and her original bag-making machine is in the Smithsonian Museum in Washington DC.

STEPHANIE KWOLEK

INVENTING THE BULLETPROOF VEST
INVENTION: KEVLAR

Law enforcers test a new bulletproof vest in 1923 before the days of Kevlar, Washington DC.

Born in New Kensington, Pennsylvania, in 1923, Stephanie Kwolek inherited her love of fabrics and sewing from her mother. She also had a fascination for science and graduated from Margaret Morrison Carnegie College in 1946 with a degree in chemistry. She found a job with the chemical giant DuPont, when few women were hired in industrial chemistry, developing low-temperature processes for finding petroleum-based fibers that displayed extraordinary strength and rigidity, fibers that could stand up to the most extreme of conditions.

After researching for nine years, she discovered that in certain circumstances, large numbers of polyamide molecules lined up in parallel, forming cloudy, liquid crystalline solutions. The liquid was "cloudy, opalescent upon being stirred, and of low viscosity" and this cloudiness might have persuaded any researcher to throw the mixture away and start again, but Kwolek had an enquiring mind and decided to persevere with it.

Initially, the man in charge of the spinning process, Charles Smullen, refused to spin this substance, fearing that it contained particles that would block the tiny holes in the spinneret. Finally, he relented and the result was fibers that were stronger and more rigid than any that had been created up to that point. When she took it to her supervisor and then the lab director, they realized that this accidental discovery was something special and that, in fact, it opened up an entirely new field of polymer chemistry. The potential for such a resilient fiber was enormous. It could withstand bullets, extremes in temperature, and it would not tear in any circumstances.

Kevlar, as the material was named, is a plastic with a very high tensile strength. It is very lightweight, stiff, and, pound for pound, five times stronger than steel! Kevlar is now often used in the field of cryogenics, is also chemical- and flame-resistant and has gone on to save countless lives through its use in bulletproof vests. It has many other uses including in tires, as underwater optical-fiber cabling, as super-strong ropes in suspension bridges and for canoes, drumheads, audio equipment, sportswear, and frying pans.

In 1995 Stephanie Kwolek became the fourth woman to be added to the National Inventors Hall of Fame. She passed away on June 18, 2014, at the age of 90.

HEDY LAMARR

FREQUENCY-HOPPING WIRELESS TECHNOLOGY
INVENTION: SECRET COMMUNICATION SYSTEM

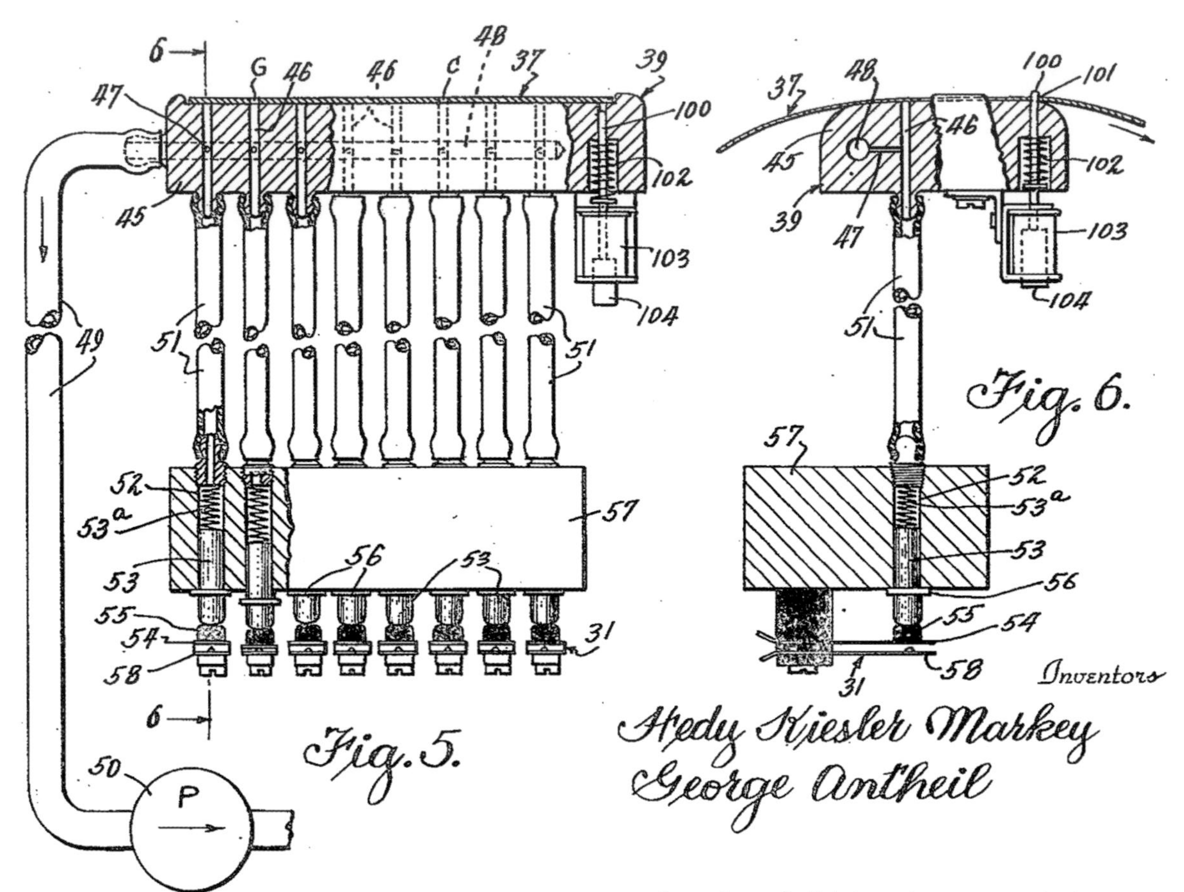

Patent diagram for Hedy Lamarr's secret communication system taken out in her married name of Hedy Kiesler Markey.

When Hollywood superstar Hedy Lamarr died in 2000, many were surprised to learn that not only was she one of the silver screen's all-time greats, she was also an inventor. During World War II, in partnership with composer George Antheil, she invented a guidance system that was intended to make radio-guided torpedoes harder for enemies to detect or to jam. This early version of frequency-hopping was designed to switch across eighty-eight different frequencies, and Lamarr and Antheil took out US Patent 2,292,387 for their Secret Communication System.

Sadly, there is no record of a working device ever being produced and the military did not adopt it during the war. However, they did begin to use it in the 1960s and the basic principles of Lamarr and Antheil's invention have now been applied to Wi-Fi and Bluetooth technology.

THE ECSTASY LADY

Hedy Lamarr was born into a Jewish family in Austria in 1914 as Hedwig Eva Maria Kiesler, and launched her acting career in the late 1920s. Initially working as a script girl, she first starred in Gustav Machaty's controversial film, *Ecstasy*, in which her face was shown mid-orgasm and in which she was also seen naked, albeit briefly. The film became a global sensation, treated in Europe very much as an art film, in the United States it was viewed as salacious and received nothing but negative publicity.

Eva was disillusioned with the film industry after *Ecstasy* and claimed that she had been tricked by the director into playing those scenes. She returned to the stage and began to pick up plaudits and gained many admirers, one of whom was Friedrich Mandl, a 33-year-old arms manufacturer, reputed to be the third-richest man in Austria.

They fell in love and, at age 18, Eva married Mandl. She soon discovered him to be controlling and domineering, and he insisted she give up her acting career. She began attending business meetings with him, learning about military technology from his discussions with scientists and engineers.

THE WORLD'S MOST BEAUTIFUL WOMAN

Finding the marriage suffocating, she eventually left him, fleeing first to Paris, disguised as her maid. She had cleverly persuaded her husband to allow her to wear all of her jewelry to a party and then disappeared immediately afterward, taking the jewelry with her.

In Paris in 1937, she was introduced to Louis B. Mayer, head of Metro-Goldwyn-Mayer, who persuaded her that she should change her name to Hedy Lamarr to negate her connection to *Ecstasy*, especially as she was known everywhere as "The Ecstasy Lady." The following year he took her to Hollywood where he promoted her as "the world's most beautiful woman."

SULTRY SEDUCTRESS

The first film in which she featured was *Algiers*, playing opposite the French actor Charles Boyer. She was an immediate sensation and Mayer began to envisage her as the new Greta Garbo or Marlene Dietrich. Normally cast as a sultry, exotic seductress, she starred opposite many of Hollywood's greatest stars in the next two decades including Clark Gable and Spencer Tracy.

But her off-screen life was far from glamorous and she was often homesick and lonely, shunning the parties and the adoring fans. Instead, her spare time was spent working on various inventions. These included an improved traffic light and a tablet that, when dissolved in water, made a carbonated drink. One man who supported her in this work was the rich industrialist and film producer, Howard Hughes. He allowed her to use his scientific engineers, and he often came to her for advice on technical matters.

SERIOUSLY FOCUSED INVENTOR

After World War II started, Lamarr learned that the enemy was able to jam Allied torpedoes, forcing them to miss their target. She had gained an understanding of torpedoes from Friedrich Mandl and reasoned that a frequency-hopping signal could prevent the jamming. A broadcast signal carrying commands to the torpedo could be bounced from frequency to frequency at split-second intervals, meaning that the broadcast could not be interfered with or jammed.

She worked with George Antheil on the idea and they were granted a patent in August 1942 for their Secret Communication System, Lamarr under her then married name, Hedy Kiesler Markey. Although it was not used during the war, it was an important invention. The technology they devised is now the principal anti-jamming method in defense systems today.

Lamarr was very focused on inventing and seriously considered giving up acting. Nonetheless, she continued in movies until 1958, and she made her last film when she was only 44 years old. As she got older, she became increasingly reclusive, retiring eventually to Miami Beach in Florida. She died in 2000, at age 85. Her ashes were interred in the Vienna Woods in Austria.

WHAT IS FREQUENCY-HOPPING?

Frequency-hopping spread spectrum (FHSS) is a wireless technology that spreads its signal over rapidly changing frequencies. Each available frequency band is divided into sub-frequencies. Signals rapidly change ("hop") among these in a predetermined order. Interference at a specific frequency will only affect the signal during that short interval. FHSS can, however, cause interference with adjacent direct-sequence spread spectrum (DSSS) systems. A sub-type of FHSS used in Bluetooth wireless data transfer is adaptive frequency-hopping spread spectrum (AFH). The earliest mentions of frequency-hopping in scientific papers and literature are in US patent 725,605 awarded to Nikola Tesla in March 17, 1903.

A turn signal of a vintage car from the 1920s. The bright red lever-like arm known as a "trafficator" flipped out to indicate that the car was about to turn.

FLORENCE LAWRENCE

THE STAR WHO CHANGED DIRECTION
INVENTION: AUTOMOBILE TURN SIGNALS

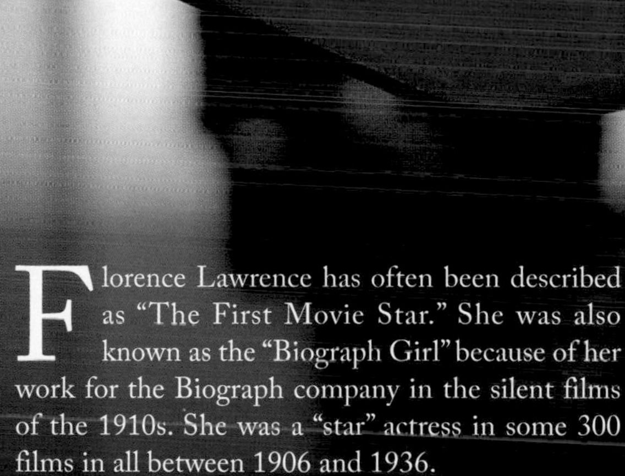

Florence Lawrence has often been described as "The First Movie Star." She was also known as the "Biograph Girl" because of her work for the Biograph company in the silent films of the 1910s. She was a "star" actress in some 300 films in all between 1906 and 1936.

Born in Hamilton, Ontario, in 1886, Florence Annie Bridgwood was usually known as "Flo." She was the daughter of Charlotte Bridgwood, the stage actress and inventor of the automatic windshield wiper. Flo was an inveterate car enthusiast and was the first person to develop "turn signals" for a car. It was a primitive system but it worked and led to the blinking turn signal lights of today.

Concerned that no one knew if she was about to make a turn which increased the possibility of an accident, Flo had flags fitted to each side of her car and when she was about to make a right or left turn, the appropriate flag was raised by her to indicate her vehicle's movement and the direction in which it was going. She also developed a simple means of warning other drivers that she was stopping, holding a sign saying "Stop" out of her driver's side window.

It is incredible to think that, although, of course now a legal requirement, turn signals had not been thought of up to that point. Florence Lawrence never patented either of these ideas but she paved the way for their later development. On December 28, 1938, Lawrence was found unconscious in her West Hollywood apartment; she had killed herself by eating ant paste.

SONJA DE LENNART

CREATING A TIMELESS CLASSIC

INVENTION: THE ORIGINAL CAPRI PANTS

Audrey Hepburn wearing striped Capri pants and a white blouse in the 1950s.

Capri pants, the pants cropped below the knee, were invented in 1948 by the fashion designer Sonja de Lennart who was born in Breslau, Prussia (now Poland), in 1920. De Lennart had always wanted to be a fashion designer and, against the wishes of her industrialist father, she studied design, secretly working as an apprentice at the Erich Boehm Atelier and later at the Herman Palm Atelier in Berlin. She learned tailoring and finally graduated as a textile engineer.

SALON SONJA

After World War II, she opened her first boutique, "Salon Sonja" in Munich, but her big break arrived when she was given a rare opportunity to showcase her designs at a leading fashion fair, Handwerksmesse. Lines formed at her stand as people clamored for her clothes. She added new designs and exhibited at other fairs and her business took off. She designed leather vests, three-quarter-length coats, and a wide-swinging skirt with a wide belt, all sold under the brand name "Capri Collection," the name coming from her family's love of the Italian island of Capri. It also came from her favorite song, "Isle of Capri."

She created Capri pants in 1948. For years, women had been wearing pants but they were usually wide and slightly masculine in aspect. De Lennart created three-quarter-length pants that had a short slit on the outer side of each leg and they came in specific lengths for summer and winter wear.

STARTING A FASHION REVOLUTION

In 1952, Edith Head, the great Hollywood costume designer put Audrey Hepburn into Capri pants in the 1953 movie *Roman Holiday*. They were accompanied by a Capri blouse and the wide Capri belt. She also wore them in *Sabrina*. Moviegoers began to seek them out and other actresses began to wear them, stars such as Brigitte Bardot, Doris Day, Jane Russell, Katherine Hepburn, Ava Gardner, Elizabeth Taylor, Marilyn Monroe, and even First Lady Jacqueline Kennedy. Women in Paris copied the singer Juliette Greco, wearing black roll-neck sweaters and flat sandals to go with their Capri pants.

In 1955, de Lennart opened a new fashion house, calling it "Maison Haase" and her designs were sought after by movie stars, aristocrats, opera directors, and Hollywood costume designers. A leading pioneer of her time, Sonja de Lennart influenced many of the great post-war designers. Capri pants became a timeless classic and featured in the collections of a whole generation at some point. In 2017, Sonja reached the grand old age of 97, and holds a special place in fashion's exclusive hierarchy.

Brigitte Bardot in polka-dot Capri pants in 1959. She was one of the best known sex symbols of the 1950s and 1960s and was widely referred to simply by her initials, BB.

THE ISLAND OF CAPRI

Capri, the location that inspired Sonja de Lennart's designs, is a rocky island in Italy's Bay of Naples, famed for its rugged landscape, upscale hotels, and shopping, anything from designer fashions to limoncello and handmade leather sandals. The sides of the island are perpendicular cliffs and the surface is composed of even more cliffs. One of its best-known natural sites is the Blue Grotto, a dark cavern where the sea glows electric blue, the result of sunlight passing through an underwater cave. In summer, Capri's dramatic, cove-studded coastline draws many luxury yachts. The island has a mythical charm, chronicled through the writings and legends of Ancient Greece. In the works of Homer, the hero Odysseus sailed past the island, narrowly escaping the fate of those who hear the voices of the Sirens.

The island of Capri seen from the sea.

RITA LEVI-MONTALCINI

THE NEUROBIOLOGY OF CHICKEN EMBRYOS
INVENTION: NERVE-GROWTH FACTOR

Nobel laureate Rita Levi-Montalcini (1909 – 2012) is the first winner of the prestigious award ever to reach the age of 100 which she did on April 22, 2009. She was born in Turin, Italy, into a family of Sephardic Jews. Her father, an electrical engineer, was also a gifted mathematician and her mother was a talented painter. Her brother became a well-known architect. Rita's father was an old-fashioned man and decided that his three daughters should not have much of an education, and that they certainly would not be allowed to go to university.

FLEEING THE GERMAN ARMY

Unwilling to fulfill the destiny her father had in mind for her, Rita pleaded with him to allow her to have a profession. Finally, he gave his reluctant permission and she began working to fill in the gaps in her education. She graduated from high school and enrolled at medical school in Turin where she studied biology under the illustrious Italian histologist, Giuseppe Levi. Two other scientists from that class went on to win Nobel Prizes—Salvador Luria and Renato Dulbecco.

Graduating with honors in 1936, she enrolled in courses in neurology and psychiatry, at this point still unsure whether she would go into medicine or commit herself to research in neurology. It all became academic, however, when that year Mussolini issued the "Manifesto per la Difesa della Razza," his racial manifesto, followed by laws that banned non-Aryan citizens from having academic or professional careers. Rita left Italy and worked for a brief period at a neurological institute in Brussels but had to leave on the eve of the invasion of Belgium by the German Army.

FACING DEATH ON A DAILY BASIS

Returning to Turin, she found her family trying to decide whether to flee to America or to find employment in accordance with the new race laws. They decided to stay and Rita built a small research unit in her bedroom. Before long, she was joined in her work by Giuseppe Levi who had also fled Belgium to escape the Nazis. Her work was inspired by "a 1934 article by Viktor Hamburger reporting on the effects of limb extirpation in chick embryos." She began researching the growth of nerve fibers in chicken embryos.

In 1941, Turin came under heavy air attack by the Allies, forcing Rita and Levi to relocate their work to the countryside but this lasted only until 1943 when Germany invaded Italy. They fled once more, this time to Florence, and lived in hiding for the remainder of the war. In August 1944, when the Allies forced the Germans out of Florence, Rita was taken on as a medical doctor at a refugee camp where she had to deal with epidemics of typhus and other infectious diseases, facing death on a daily basis alongside the refugees.

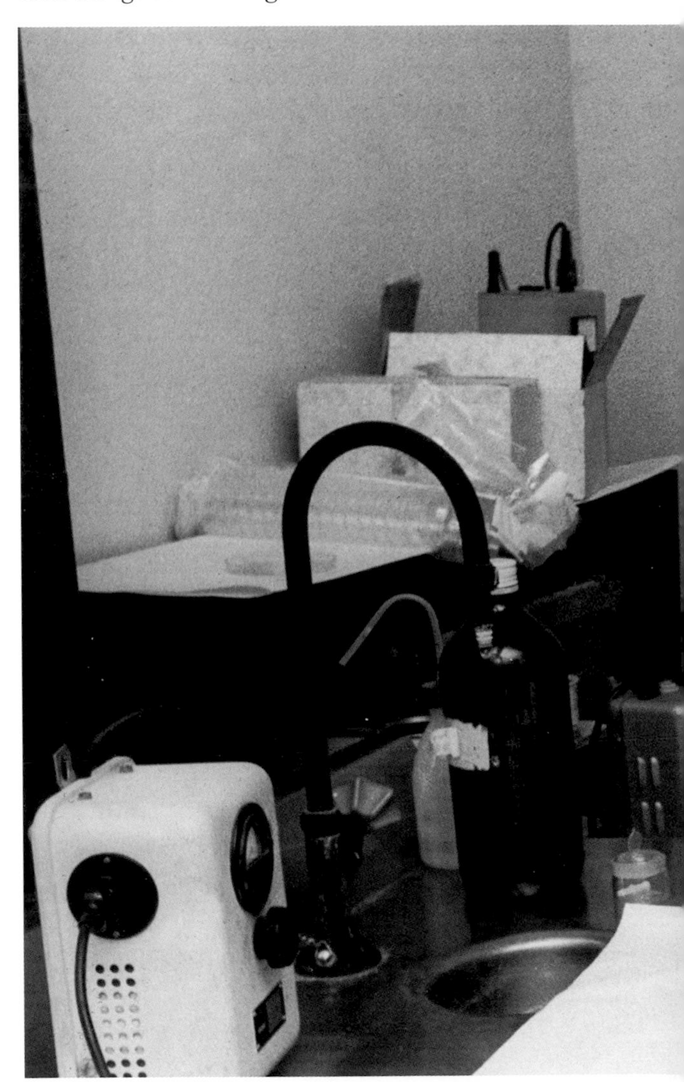

NERVE-GROWTH FACTOR

When the war finally ended in May 1945, she returned to Turin with her family and resumed her old positions at the university. In the fall of 1947, she was invited by Professor Viktor Hamburger to go to Washington University in St. Louis, Missouri, and repeat the experiments on chicken embryos that she had carried out years earlier. Her work was to isolate nerve-growth factor by observing certain cancerous tissues that cause extremely rapid growth of nerve cells.

Transferring segments of tumors to chicken embryos, she established a mass of cells full of nerve fibers. She was surprised to discover that nerves were growing everywhere like a halo around the tumor cells. When describing it, Montalcini said it was "like rivulets of water flowing steadily over a bed of stones." She deduced that the tumor was itself releasing a substance that was stimulating the nerve growth. This was nerve-growth factor.

MAST CELLS IN HUMAN PATHOLOGY

She had planned to be in St. Louis for just ten or twelve months but she stayed for 30 years, and in 1956 was made Associate Professor, and in 1958 Full Professor. In 1962, she established a second laboratory in Rome and divided her time between there and St. Louis.

After she retired in 1977, she was appointed as director of the Institute of Cell Biology of the Italian National Council of Research in Rome. She was awarded the 1986 Nobel Prize in Physiology or Medicine jointly with colleague Stanley Cohen for the discovery of nerve-growth factor. In the 1990s, she was one of the first scientists to point out the importance of mast cells in human pathology.

She died in her home in Rome on December 30, 2012, at the age of 103. Upon her death, the Mayor of Rome, Gianni Alemanno, stated it was a great loss "for all of humanity."

Nobel Prize winner Professor Rita Levi-Montalcini at the Washington University School of Medicine, St. Louis, Missouri.

ADA LOVELACE

THE ANALYTICAL COUNTESS
INVENTION: COMPUTER PROGRAMMING

The remarkable Ada Lovelace was born in 1815, the only child of the poet Lord Byron and his wife Anne Isabella Milbanke, Lady Wentworth. Byron had other children, but Ada was the only legitimate one. A month after her birth, the poet and his wife separated and in 1816 he left England for ever. He died in 1824 during the Greek War of Independence when Ada was 8 years old. Lady Wentworth, known as Annabella, encouraged her daughter into the study of mathematics and logic because of the bitterness she felt toward her husband.

Ada received an excellent education: she was tutored in mathematics by the science writer and polymath Mary Somerville, and by logician and mathematician Augustus De Morgan. Under De Morgan's tutelage, Ada became well schooled in the principles of algebra, logic, and calculus. Her mathematical education was unusual at the time, even for someone from her wealthy background. While mathematics flourished in continental Europe in the first half of the nineteenth century, British mathematics was floundering.

ANALYST IN POETICAL SCIENCE

As a child, Ada was often ill and around the time of her father's death, she began to suffer from headaches and loss of vision. As a result, she was confined to bed for a year.

In 1828, when she was 12, she became determined to learn how to fly. To this end, she constructed a pair of wings, having carefully analyzed the anatomy of birds to achieve the correct proportions and distances between the wings and the body. She even wrote a book—*Flyology*—detailing her research into flying. In 1835, she married William King who was made Earl of Lovelace three years later. She became Countess of Lovelace and they had three children.

Ada liked to describe her approach to her work as "poetical science," portraying herself as an "Analyst (& Metaphysician)." She must have been inspired by some of the great scientists she encountered in her social life, men such as the pioneer of electricity Andrew Crosse, physicist and mathematician Sir David Brewster, scientist Charles Wheatstone, and Michael Faraday, one of the greatest ever British scientists.

MEETING CHARLES BABBAGE

Her most important working relationship was the long one she enjoyed with another mathematician, Charles Babbage, often described today as the "father of computer." She worked with Babbage on his plans for the Analytical Engine, a mechanical general-purpose computer that he was designing. Ada first met Babbage in June 1833, introduced by Mary Somerville. The Italian mathematician Luigi Federico Menabrea, had written an article on Babbage's proposed Analytical Engine that Ada translated for him between 1842 and 1843, adding her own set of notes regarding the engine.

Titled simply *Notes*, these contain what is considered to be the first computer program, an algorithm that was designed to be run by a machine. Sadly, as the Analytical Engine was never actually built, her program was never tested. Ada was something of a computer visionary, predicting the future with her belief that computers could do more than calculating or number-crunching. Her *Notes* end with her program for computing Bernoulli numbers. It was far more ambitious and complex than any program Babbage had written for the Engine.

PIONEERING WOMAN IN A MAN'S WORLD

After working with Babbage, Ada undertook various projects. She became interested in the brain and its workings, telling a friend of her desire to create a mathematical model of how the brain and nerves work, "a calculus of the nervous system" as she described it. It never developed into anything. Ada Lovelace died of cancer in 1852, at age just 36, the same age as her father Lord Byron, when he passed away.

THE ANALYTICAL ENGINE

Charles Babbage's Analytical Engine was never completed, but a portion (above) consisting of part of the mill (CPU) and the printing devices was assembled shortly before his death in 1871. The engine would have been programmed using punched cards, an idea borrowed from the Jacquard loom for weaving patterned cloth. Ada Lovelace wrote, "We say most aptly that the Analytical Engine weaves algebraical patterns just as the Jacquard-loom weaves flowers and leaves." Cards were a particularly clever solution for weaving cloth, or for performing calculations, because they allowed any desired pattern, or equation, to be generated automatically.

LIZZIE MAGIE

GET OUT OF JAIL FREE!
INVENTION: THE LANDLORD'S GAME

Monopoly, the classic real estate trading game from Parker Brothers was first introduced to America in 1935.

Created in 1932 by domestic heater salesman Charles Darrow, *Monopoly* is played all over the world. But when it was invented there were a number of similar board games including one called *Finance* or *The Fascinating Game of Finance*. The precursor to all these variants on the same idea of moving round a board purchasing property and utilities, was a game called *The Landlord's Game*, designed by American Elizabeth Magie (1866 – 1948). The game was first patented by her in 1904 although it was certainly being played as early as 1902.

Magie was a follower of the popular American political economist Henry George (1839 – 97) whose philosophy was known as Georgism. It was based on a belief that people should own the value they themselves produce but that the economic value of land should be owned equally by each member of society. Magie created *The Landlord's Game* to educate players about the principles of Georgism, or as she put it, a "practical demonstration of the present system of land grabbing with all its usual outcomes and consequences."

In 1906 Magie and some fellow Georgists formed a company, the Economic Game Company, to publish her game but without much success. Three years later, she invited Parker Brothers, eventual publishers of *Monopoly*, to publish the game and another she had devised, but although they did publish the other game, they rejected *The Landlord's Game*, insisting that it was too complicated for a board game.

Magie held the patent until 1935 when she sold it for $500 to Parker Brothers who had recently begun selling *Monopoly*. Charles Darrow claimed to have invented *Monopoly* although it was obvious that it had been plagiarized from *The Landlord's Game*. They printed a few copies of Magie's game to secure the copyright, and those examples are now among the rarest and most collectible of all twentieth century board games.

(Facing page) The similarities between Lizzie Magie's *The Landlord's Game* and *Monopoly* are plain to see.

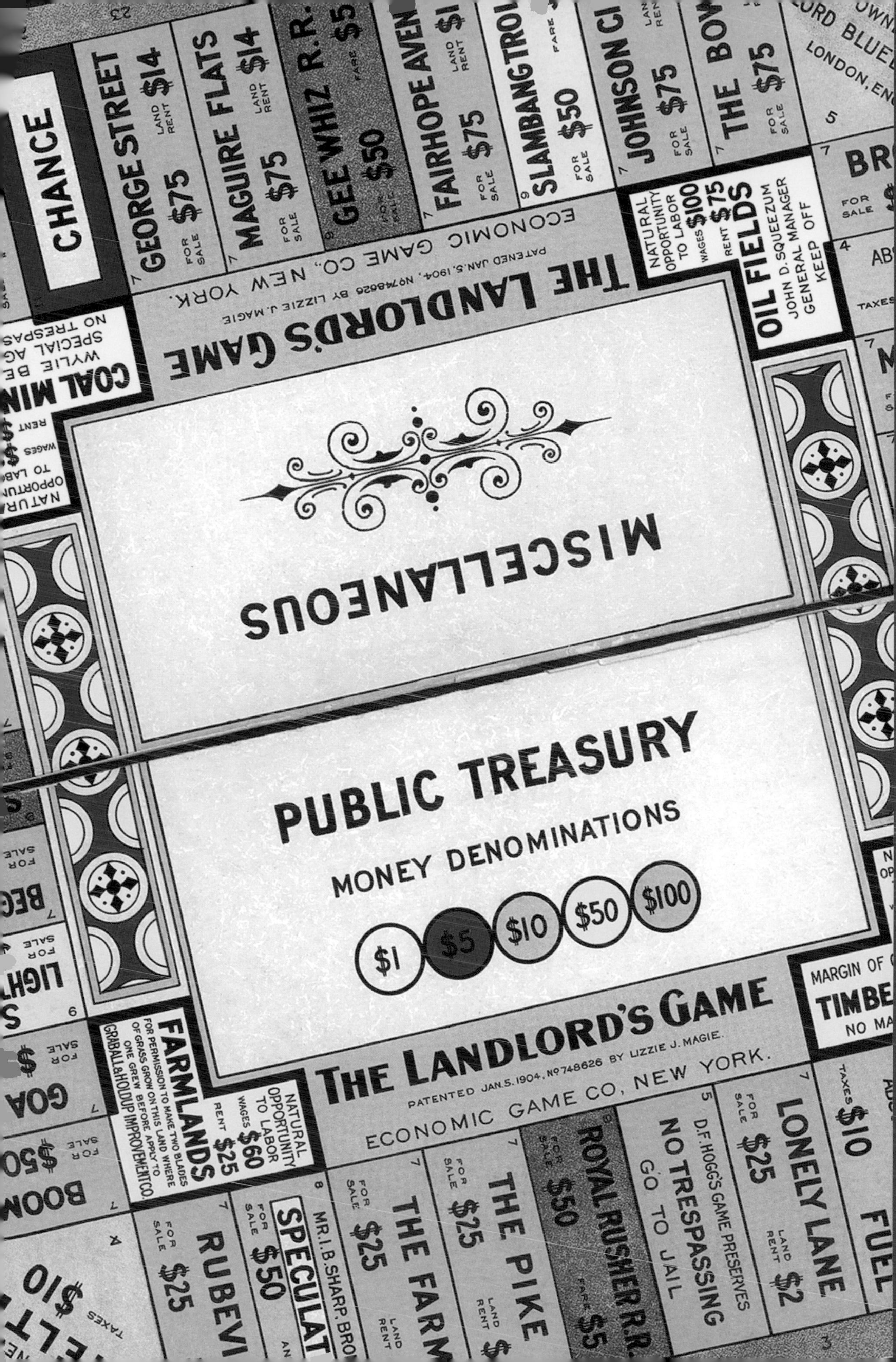
THE LANDLORD'S GAME
PATENTED JAN. 5. 1904, No 748626 BY LIZZIE J. MAGIE.
ECONOMIC GAME CO, NEW YORK.
MISCELLANEOUS
PUBLIC TREASURY
MONEY DENOMINATIONS
$1
$5
$10
$50
$100
CHANCE
GEORGE STREET
MAGUIRE FLATS
GEE WHIZ R.R.
FAIRHOPE AVEN
SLAMBANG TROL
JOHNSON CI
OIL FIELDS
JOHN D. SQUEEZUM
GENERAL MANAGER
KEEP OFF
NATURAL OPPORTUNITY TO LABOR
FARMLANDS
FOR PERMISSION TO MAKE TWO BLADES OF GRASS GROW ON THIS LAND WHERE ONE GREW BEFORE APPLY TO GRABALL & HOLDUP IMPROVEMENT CO.
SPECULAT
MR. I. B. SHARP, BRO
THE FARM
THE PIKE
ROYAL RUSHER R.R.
NO TRESPASSING
GO TO JAIL
D.F. HOGG'S GAME PRESERVES
LONELY LANE
FUEL
MARGIN OF
TIMBE
COAL MIN
FOR SALE
LAND RENT
TAXES

JOY MANGANO

INVENTING JOY

INVENTION: THE MIRACLE MOP

Joy Mangano selling her Miracle Mop on QVC.

Inventors these days often also have to be shrewd self-marketeers, writing books, appearing on television shows and shopping channels to push their wares. Joy Mangano is one of the best self-marketeers in the business. She appears regularly on the US shopping channel, Home Shopping Network, and has written a book about living a successful life, *Inventing Joy* which has the subtitle *Dare to Build a Brave and Creative Life*. She has become famous and actress Jennifer Lawrence was nominated for an Academy Award for her portrayal of her in the 2015 biopic *Joy*.

Born in East Meadow, New York, in 1956, Mangano graduated in 1978 from Pace University with a degree in business administration. In 1990, fed up with the mops she was using, she dreamed up the Miracle Mop, a self-wringing plastic mop the head of which was a continuous loop, about 300 feet (90 meters) long, made of cotton. It could be wrung out with a twist of the handle, meaning that the user did not have to get his or her hands wet.

Borrowing money to create a prototype, she produced a hundred mops that she sold at trade shows and locally where she lived on Long Island. She had another thousand made and sent them to the television shopping channel QVC. Initial sales were not impressive but when Mangano herself appeared on the channel selling the mop, sales rocketed and she sold a staggering 18,000 in less than thirty minutes. By the year 2000, sales of Miracle Mops were bringing in an astonishing $10 million a year.

Joy Mangano is now Home Shopping Network's most successful salesperson with annual sales of more than $150 million.

Movie poster for the 2015 biopic *Joy*, starring Jennifer Lawrence.

AGNES MARSHALL

FREEZING OF THE ICE QUEEN

INVENTION: THE ICE CREAM FREEZER

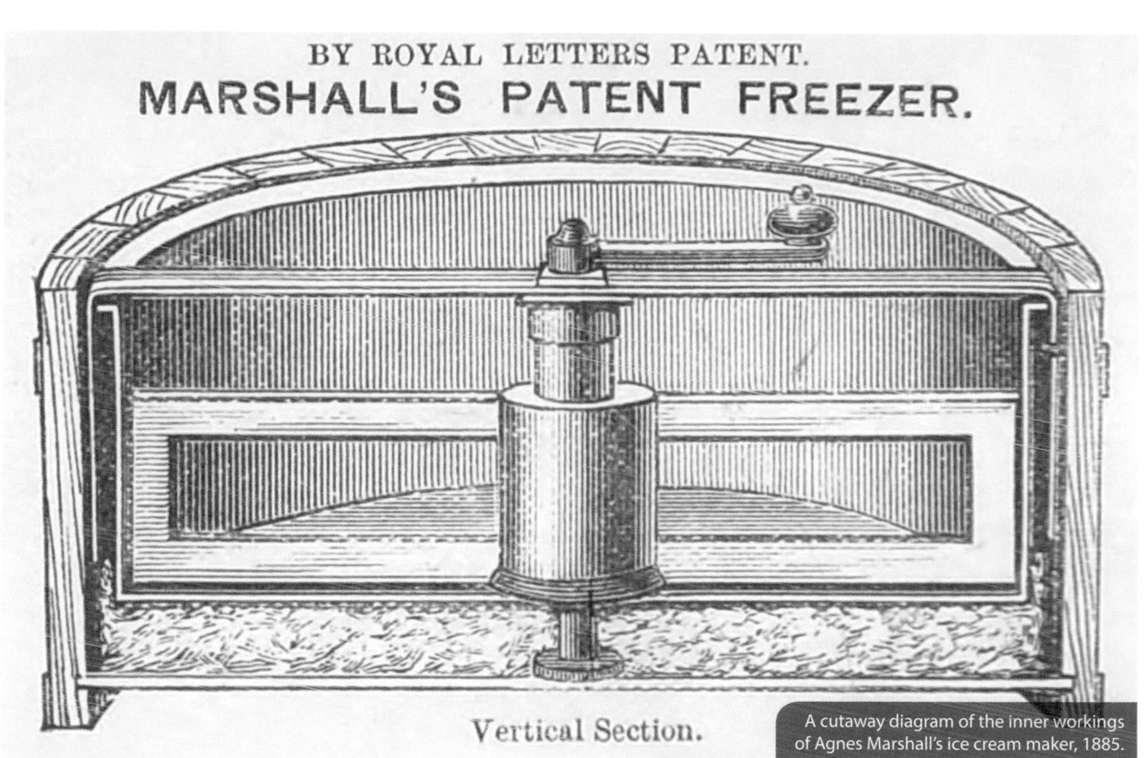

A cutaway diagram of the inner workings of Agnes Marshall's ice cream maker, 1885.

Agnes Marshall was a celebrity chef in the nineteenth century, a leading cookery writer of the time who was known as "Queen of Ices" because of her books on ice cream and frozen desserts. She has been credited with inventing in 1888 the first portable, edible ice cream cone called a "cornet" which was made from ground almonds.

Born in Walthamstow, London, in 1855, Agnes Marshall studied cooking from an early age and may even have trained in Paris and Vienna. An enterprising lady, she published four cookery books, lectured about cooking, and established an agency that supplied domestic staff.

In 1883, she and her husband purchased the National Training School of Cookery in Mortimer Street, London, and renamed it the Marshall School of Cookery. They are reputed to have had almost 2,000 students within a few years, taught by specialists. They also opened a food shop and a household equipment shop. Flavorings, spices, and syrups were all sold under the Marshall own-brand name.

Knowing so much about ice cream, as her books testified, it is no surprise that she invented a hand-cranked ice cream freezer which, it was claimed, could freeze a pint of ice cream in less than five minutes. An advertisement for "Marshall's Patent Freezer" boasted:

> *Marshall's Patent Freezer is praised by all who know it for cheapness in first cost, cleanliness in working, economy in use, and simplicity in construction. Rapidity in freezing. No packing necessary. No spatula necessary. Smooth and delicious ice produced in 3 minutes.*

After her death in 1905, the rights to her cookbooks were sold to the publisher of Mrs. Beeton's cookbooks, Ward Lock. Agnes's husband took over the business but without her drive and personality it failed and her name faded into obscurity.

MARIE MARVINGT

LANDING ON THE SANDS OF THE SAHARA
INVENTION: METAL SKIS ON AIR AMBULANCES

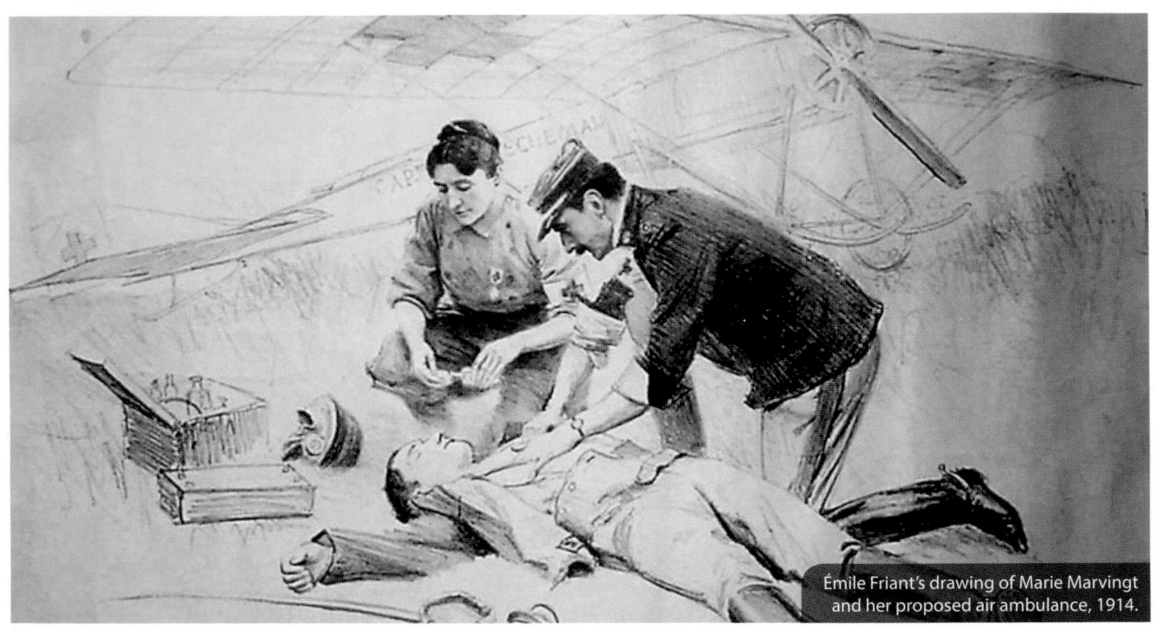

Émile Friant's drawing of Marie Marvingt and her proposed air ambulance, 1914.

Marie Marvingt (1875 – 1963) was a Wonder Woman. As a child she was encouraged by her father to take part in sports, and by the age of 5 she could swim 4,000 meters. In 1890, at the age of 15, she canoed 400 kilometers from her home in Nancy in France to Koblenz in Germany. She swam, cycled, climbed mountains, skied, flew planes, rode horses, participated in athletics, shot rifles, and fenced.

In winter sports she reigned supreme between 1908 and 1910, coming first in competition twenty times and winning the women's world bobsleigh championship in 1910. Refused entry to the Tour de France in 1908 due to her gender, she rode the course anyway. In 1910, the French Academy of Sports awarded her a gold medal for all sports, an award that has never been granted to anyone else, man or woman.

But aviation was her passion, beginning with hot-air balloons and then moving on to aircraft. In 1909, she was the first woman to pilot a balloon across the North Sea and English Channel from France to England. She was the first woman to fly solo in a monoplane and became a much anticipated star of many air shows.

In World War I, she disguised herself as a man, served at the front, and took part in operations with a regiment of Italian Alpine troops in the Dolomite Mountains. She volunteered to fly bombing missions over German-held territory and was awarded the Croix de Guerre for her bombing of a German base in Metz.

For much of the remainder of her life Marie Marvingt was devoted to medical evacuation by air ambulance, giving lectures at more than 3,000 conferences around the world. She also worked as a journalist, a war correspondent, and was a medical officer with French forces in North Africa. It was around this time that she came up with the idea of metal skis for air ambulances working in Morocco and Algeria. The skis enabled aircraft to land on the sands of the Sahara. For good measure, she ran a ski school in the desert, teaching the locals to ski on the sand dunes.

During World War II, Marvingt established a convalescent center for wounded aviators. She also served as a surgical nurse and found time to invent a new type of surgical suture. In 1955, at age 80, she learned to fly a helicopter and flew faster than the speed of sound in a US F101 Voodoo jet fighter plane. Marie Marvingt died in 1963, at age 88, the most decorated woman in French history and, undoubtedly, one of the most remarkable.

MARY THE JEWESS

THE FIRST WESTERN ALCHEMIST
INVENTION: THE ALEMBIC AND THE BAIN-MARIE

Alchemists using a bain-marie, from illustration published in *Bibliotheca Chemica Curiosa*, 1702.

Mary the Jewess was an alchemist who featured in the works of the Gnostic Christian writer, Zosimos of Panopolis. She wrote what are the oldest existing books on the subject of alchemy and was probably the first real alchemist in the West. Mary, who lived sometime between the first and third centuries, is described by Zosimos as "one of the true sages." Her written works have not survived but some quotations, supposedly by her, have been found. In *The Dialogue of Mary and Aros on the Magistery of Hermes*, she is reported to have instructed:

> *Keep the fume and take care that none of it fly away. And let your measure be with a gentle fire such as is the Measure of the heat of the Sun in the Month of June or July.*

Mary is said to have created what became known as the "Axiom of Maria" which stipulates: "One becomes two, two becomes three and out of the third comes the one as the fourth." It was an axiom borrowed by the Swiss psychiatrist Carl Jung (1875 – 1961) in his concept of individuation.

A number of inventions of chemical instruments have been attributed to Mary, that are still used today:

- The Tribikos is an alembic with three arms that is used to obtain substances that have been purified by distillation.
- The Kerotakis is an airtight container used in the hermetic arts and gives us the phrase "hermetically sealed."
- More famously Mary's name lives on in the cooking term "bain-marie" which means "Mary's bath." In cooking this double boiler is used to melt chocolate without splitting and to prevent a crust forming on custard.

A modern day still-spirits alembic.

SYBILLA MASTERS

THE FIRST WOMAN INVENTOR IN AMERICAN HISTORY

INVENTION: THE CORN MILL

The first mention of Sybilla Righton (*c.* 1676 – 1720) is in the records of the New Jersey colony for 1692 when she appeared as a witness for her father in court. Sometime between 1693 and 1696, she married a wealthy Quaker merchant and landowner, Thomas Masters, and the couple went on to have four children. They moved to Philadelphia where Thomas served two terms as Mayor of Philadelphia in 1707 and 1708.

At that time one of the most common foods was ground-up Indian corn that was called hominy grits. Producing it was hard work as the corn was ground between two large millstones. Observing the local Native American women, Sybilla saw they used large wooden posts to pound the corn instead of stones. Taking this as her inspiration, she invented a mill that used hammers to pound the corn into meal that she named Tuscarora Rice after the Native American tribe. Her invention allowed people to process the corn into different foods as well as cloth products.

It was not her only invention. She also created a method of weaving straw and palmetto leaves into hats and bonnets. She wanted to get patents for her two inventions, but Pennsylvania did not at the time offer them and, she decided to travel to England to apply for her patents, leaving for London in 1712.

On arrival however, she found there was no regular procedure to apply for a patent, other than apply directly to the King and then wait. While she waited, Sybilla opened a shop in London selling her straw and palmetto leaf hats and bonnets.

In 1715 King George I at last awarded her a patent for "Cleaning and Curing the Indian Corn Growing in the Seven Colonies in America" but as women were prohibited from holding patents, Patent No. 401 was issued to her husband for "a new invention found out by Sybilla, his wife." The following year her husband was also granted Patent No. 403 for Sybilla's method of weaving hats and bonnets.

Delighted, she returned home with her patents. No other American woman was granted a patent until 1793, and it could be safely argued that she was the first woman inventor in American history.

The method used by Native American Tuscarora women of pounding corn with a wooden post inspired Sybilla Masters' corn mill invention.

BARBARA McCLINTOCK

THE TRUE SIGNIFICANCE OF A DISCOVERY

INVENTION: GENETIC TRANSPOSITION

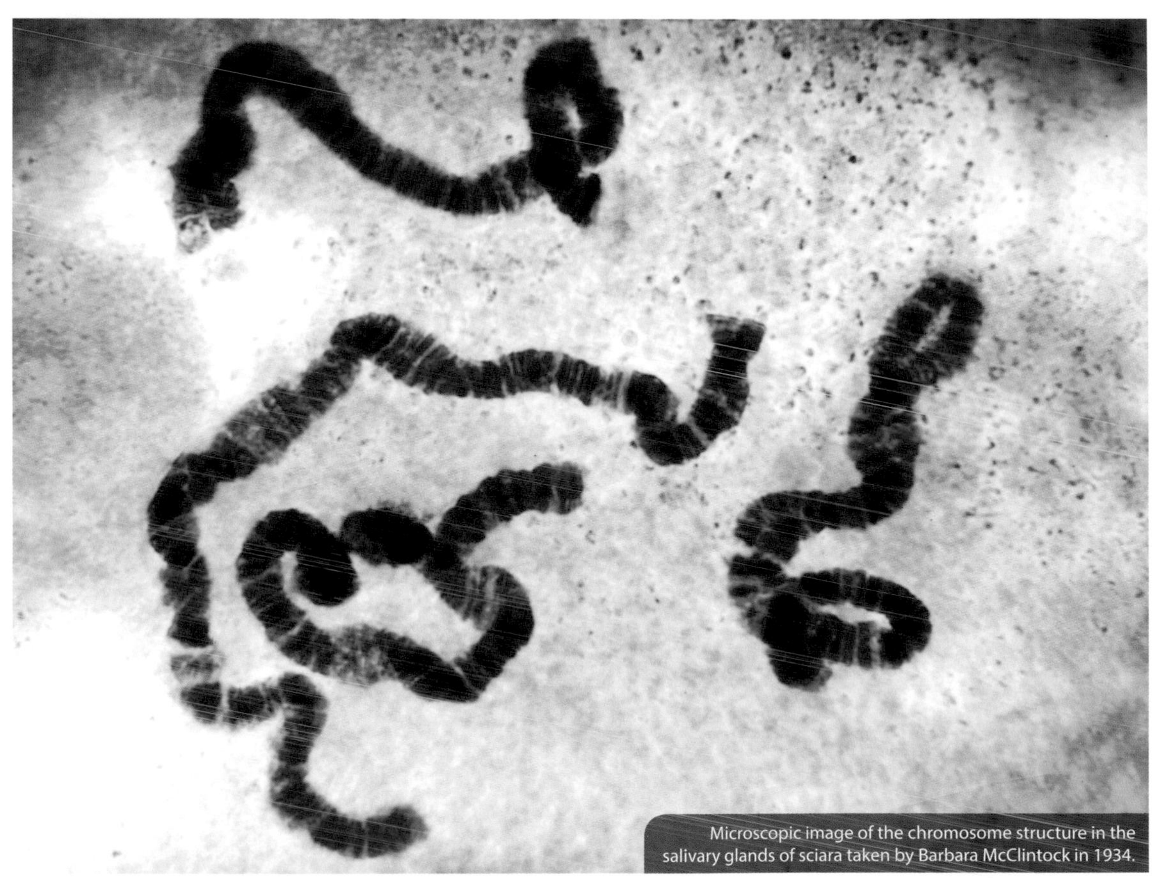

Microscopic image of the chromosome structure in the salivary glands of sciara taken by Barbara McClintock in 1934.

The Nobel Prize can take a long time to be awarded, often many years after the initial discovery or advance has been made. Barbara McClintock's award was made no less than 39 years after her discoveries. Her achievements were in 1944 and the prize was not awarded until 1983. It can often take that long for the significance of a piece of work to become apparent, sometimes it even has to wait for other scientific discoveries before its true significance is revealed. Barbara McClintock (1902 – 92) is one of twelve women to have won the Nobel Prize in Physiology or Medicine during the 116 years in which it has been awarded and she is the only woman to have received an unshared prize in her category.

Born in 1902, she was brought up in Connecticut and New York. Her background was relatively poor and her desire to undertake scientific research seemed like a fantasy to her family who would rather that she just got married so that they could get her off their hands. Nonetheless, they supported her, her father especially, and she enrolled at Cornell's College of Agriculture in 1919. In fact, her studies and research became her life and she never did marry.

She studied corn's hereditary characteristics, such as the color of the kernel, and how these are passed down through generations, linking this to changes that might happen in the plants' chromosomes. In the 1940s and 1950s, she proved that genetic elements can change position on a chromosome. When this happens it can cause adjacent genes to become active or inactive.

There was general skepticism from the scientific community toward her work, to the extent that in 1953 she stopped publishing her data. In the 1960s and 1970s, however, her research became better understood as other scientists confirmed what she had discovered all those years ago. Numerous awards ensued, including the Nobel.

LISE MEITNER

A NEW ERA IN HUMAN HISTORY
INVENTION: NUCLEAR FISSION

Lise Meitner (1878 – 1968) was an Austrian-Swedish physicist who worked on radioactivity and nuclear physics. She and Otto Hahn led the small group of scientists who first discovered nuclear fission of uranium in 1938. Their discovery eventually led to the atomic bombs which were dropped on the Japanese cities of Hiroshima and Nagasaki at the end of World War II.

Lise Meitner was born into a Jewish family in Vienna, Austria, where her father was one of the first Jewish lawyers in the country. She launched her research career early, at the age of 8 when she began to keep a notebook of her discoveries. Drawn to mathematics and science, she made studies of an oil slick, thin films, and reflected light. In 1905, she became only the second woman to gain a doctorate in physics from the University of Vienna.

WORKING WITH MAX PLANCK

With her father's financial help, she enrolled at the Friedrich-Wilhelms-Universität in Berlin, Germany, where she was delighted to be permitted to attend lectures by Max Planck, the eminent German theoretical physicist who went on to win the 1918 Nobel Prize for Physics. It was a special gesture by the great man who up to that time had refused to allow women into his lecture theaters.

A year later, she was working as Planck's assistant and she had also begun working with Otto Hahn with whom she collaborated for the next thirty years. In the early years of their collaboration, the pair discovered new isotopes (variants of a particular chemical element which differ in neutron number) and in 1909, she presented a paper on beta-radiation. Meitner moved with Hahn to the newly established Kaiser-Wilhelm-Institute in south-west Berlin where she worked as a "guest" in his Department of Radiochemistry. Only in 1913 when she was offered the opportunity to be an Associate Professor in Prague did they finally give her a permanent position.

THE GERMAN MARIE CURIE

In 1917, Meitner and Hahn discovered the first isotope of the element protactinium for which she was awarded the Leibniz Medal by the Berlin Academy of Sciences. She was put in charge of her own physics department at the Kaiser-Wilhelm-Institute. In 1926, she became the first woman in Germany to be appointed full professor of physics at Berlin University. Nine years later, she and Hahn launched what was called the "transuranium research" program that resulted in the discovery of nuclear fission of heavy nuclei in December 1938. Following this discovery, the great theoretical physicist Albert Einstein used the epithet "the German Marie Curie" to describe Meitner.

When the Nazis came to power in 1933, many Jewish scientists emigrated to the United States. Meitner, protected by her Austrian citizenship, continued working but in July 1938, she finally fled to the Netherlands, traveling under cover with only 10 German marks in her purse and the clothes she stood up in. She moved on to Stockholm where she found work in the laboratory of Swedish physicist Manne Siegbahn, although he, like Max Planck, was never happy with the idea of women scientists.

ELECTRIFYING WORLD SCIENCE

Around this time, back in Berlin, Otto Hahn and his assistant Fritz Strassmann discovered nuclear fission, a discovery, as Meitner later said, that "opened up a new era in human history." News of this event electrified scientists around the world because it was recognized that nuclear fission could be used as a weapon such as the world had never seen before.

Albert Einstein was urged to alert US President Roosevelt and as a result the Manhattan Project was initiated, leading to America developing its own nuclear weapons. Meitner was invited to work on the Manhattan project at Los Alamos but is reported to have been horrified. "I will have nothing to do with a bomb!" she insisted. After the war, she remained in Sweden and in 1947 became a professor at University College of Stockholm.

OTTO HAHN WINS THE NOBEL PRIZE

In 1945, when it was announced that Otto Hahn had been awarded the Nobel Prize for Chemistry it surprised many in the scientific community that Meitner did not share it. She was disappointed herself and expressed her feelings in a letter:

> *Surely Hahn fully deserved the Nobel Prize for chemistry. There is really no doubt about it. But I believe that Otto Robert Frisch and I contributed something not insignificant to the clarification of the process of uranium fission—how it originates and that it produces so much energy and that was something very remote to Hahn.*

After the war, she became critical of the scientists who remained in Germany and who continued with their work without protesting against Hitler and the Nazi regime. In a letter to Hahn that she never sent, she wrote:

> *... millions of innocent human beings were allowed to be murdered without any kind of protest being uttered ... first you betrayed your friends, then your children—and finally ... you betrayed Germany itself ....*

Lise Meitner died in her sleep in 1968 at the age of 89, just a few months after the man with whom she should have shared a Nobel Prize, Otto Hahn.

Photograph taken at .025 seconds after the Trinity initial detonation shows a plasma dome.

THE MANHATTAN PROJECT, LOS ALAMOS

The Manhattan Project was a research and development program led by the United States during World War II that produced the first nuclear weapons. Nuclear physicist Robert Oppenheimer was the director of the Los Alamos Laboratory, New Mexico, that designed the actual bombs. The Manhattan Project began modestly in 1939, but grew to employ more than 130,000 people and cost nearly US $2 billion (about $27 billion in present-day dollars).

Two types of atomic bombs were developed concurrently during the war: a relatively simple gun-type fission weapon called Little Boy, and a more complex implosion-type nuclear weapon called Fat Man.

The first nuclear device ever detonated was an implosion-type bomb at the Trinity test, conducted at New Mexico's Alamogordo Bombing and Gunnery Range on July 16, 1945. Little Boy and Fat Man bombs were used a month later in the atomic bombings of Hiroshima and Nagasaki. In the immediate postwar years, the Manhattan Project conducted weapons testing at Bikini Atoll as part of Operation Crossroads.

EMPRESS LEIZU

In China, ingenious women were revered as gods. To the Empress Leizu (also known as as Xi Lingshi or Hsi Ling-Shih), wife of the Yellow Emperor who ruled in the twenty-seventh century BC, is ascribed the discovery of sericulture, the cultivation of silkworms to produce silk. She is reported to have had her eureka moment when a cocoon fell into her cup while she was enjoying midday tea. The heat unwrapped the silk and it stretched across the imperial garden.

She realized that the cocoon was the source of this beautiful material. As silkworms eat mulberry leaves, she persuaded her husband to provide her with a grove of mulberry trees. Leizu is also credited with the invention of the silk reel which enabled the filaments of silk to be combined to form a thread that was robust enough to be used in weaving of silk cloth. The invention of the silk loom is also often attributed to her. For all these ancient inventions the legendary Empress Leizu is venerated in China as "Silkworm Mother."

Silk filaments being unraveled from cocoons in Xinjiang, north-west China. The ancient Silk Road trade route linking China and the Middle East passed through Xinjiang, a legacy that can still be seen today in Xinjiang's traditional open-air bazaars selling silk produce.

FLORENCE MELTON

WALKING IN ANGEL FOOTSTEPS

INVENTION: THE WASHABLE SLIPPER

You never knew you needed them, but obviously, given that they have sold in truckloads, you do! Washable slippers were accidentally invented by Florence Melton in the late 1940s while she was working on using foam latex for a women's shoulder pad she had patented. Melton was born in 1911 in Philadelphia into a family that did not have a lot of money. This meant that at age just 13, she had to go out and earn a living, working at her local branch of Woolworth's. Married at 19 to Aaron Zacks, she moved to Columbus, Ohio, and continued her retail career as a merchandiser in a department store.

In 1946, she founded a company with her husband, the R.G. Barry Corporation which grew into the world's largest supplier of comfort footwear. It was around this time that she made the discovery that foam latex was an ideal material with which to line slippers. She designed them and gave them the name Angel Treads which she later changed to Dearfoams. She then went on to sell a billion slippers which would be staggering, if you were not wearing a pair of Dearforms which provide too much foamy comfort to allow you to stagger!

Vintage Angel Treads slippers in original packaging, made by R.G. Barry Corporation in Columbus, Ohio.

ANN MOORE

SNUGGLING UP WITH AN AGE-OLD IDEA

INVENTION: SNUGLI BABY CARRIER

A simple idea, used for thousands of years by West African women—a long shawl used as a sling to strap a child securely to a woman's back so that she can work and move around but always have contact with her baby. Ann Moore saw it first-hand when she worked for the Peace Corps in Togo in West Africa in the 1960s. When she came home, she wanted to enjoy the same freedom while keeping her child close, but there was always a danger that the little one would slip out of the sling or the ends would become undone.

So, with her mother Agnes Aukerman, she developed a back harness, using a simple backpack. This became the basis of a product that she called the Snugli which she patented in 1969. It is an invention that has now been used all over the world, enabling mothers to lead an active life, ride bicycles, cook, and go shopping with their babies snugly in their Snugli.

Of her invention, Moore has said: "I didn't invent the idea. I took an age-old idea that's been going on all over the world for thousands of years and just adapted it to our Western culture." She sold the Snugli business in 1985, and then started again in 1996 with a new baby carrier for newborns called the Weego.

THE PEACE CORPS

The Peace Corps was an initiative that will be forever linked to the presidency of John F. Kennedy. He had said in 1951 while he was still in the House of Representatives: "young college graduates would find a full life in bringing technical advice and assistance to the underprivileged and backward Middle East ... In that calling, these men would follow the constructive work done by the religious missionaries in these countries over the past 100 years." A year later, his thought was echoed by Connecticut Senator Brien McMahon who proposed an army of young Americans acting as "missionaries of democracy."

Although eternally linked to Kennedy, it was Senator Hubert Humphrey who introduced the first bill on the subject in 1957. Kennedy picked up on it again during his presidential campaign in 1960 and on March 1, 1961, as president, he signed an Executive Order establishing the Peace Corps. It began recruiting in 1962 and since then around 220,000 young Americans have worked in 141 countries around the world.

Ann Moore carries her daughter on her back in 1968.

MARY SHERMAN MORGAN

SAVING THE AMERICAN SPACE PROGRAM
INVENTION: LIQUID ROCKET-FUEL HYDYNE

The NASA/ABMA Jupiter-C/Explorer 1 rocket and satellite launches from the Cape Canaveral Air Force Station, Florida, January 31, 1958.

The shortage of male scientists caused by World War II was lucky for a few women who, if the men had not gone off to fight the war, might not otherwise have found their way into a science career. Mary Sherman Morgan (1921 – 2004) was one such lady. She had been studying chemistry at Minot State University in North Dakota when America entered the war in 1941.

Learning that Morgan had a knowledge of chemistry, a local recruitment agency approached her about a top-secret job at a factory in Sandusky, Ohio. It was good money so she accepted, not quite knowing what she was getting herself in for. She arrived in Sandusky to find that the job was at the Plum Brook Ordnance Works, a munitions factory. She was responsible for the manufacture of trinitrotoluene (TNT) and other high explosives.

After the war ended, Morgan took a job at North American Aviation in California and was soon promoted to be Theoretical Performance Specialist with responsibility for calculating the anticipated performance of new rocket propellants. She was the only woman among the 900 engineers employed there and the only person without a university degree.

The German rocket scientist Wernher von Braun, then building rockets for the American space program, awarded a contract to North American Aviation to devise a more powerful fuel. Mary Morgan was named technical lead on the project as she had more experience than anyone else. She developed a fuel named Hydyne and trials were made in 1956, using Redstone rockets.

It was a hugely competitive time in the space race. The Soviets had succeeded in putting their satellite Sputnik into space and the United States' effort seemed to be in disarray. Von Braun was ordered to prepare his rocket with Morgan's fuel propelling it. Juno I launched America's first satellite—Explorer I—successfully into earth orbit on the last day of January 1958. With her discovery of a safe and workable fuel, Mary Sherman Morgan is often said to have saved the American space program.

LYDA D. NEWMAN

NO HAIR-BRAINED INVENTION!

INVENTION: THE CLEANABLE HAIRBRUSH

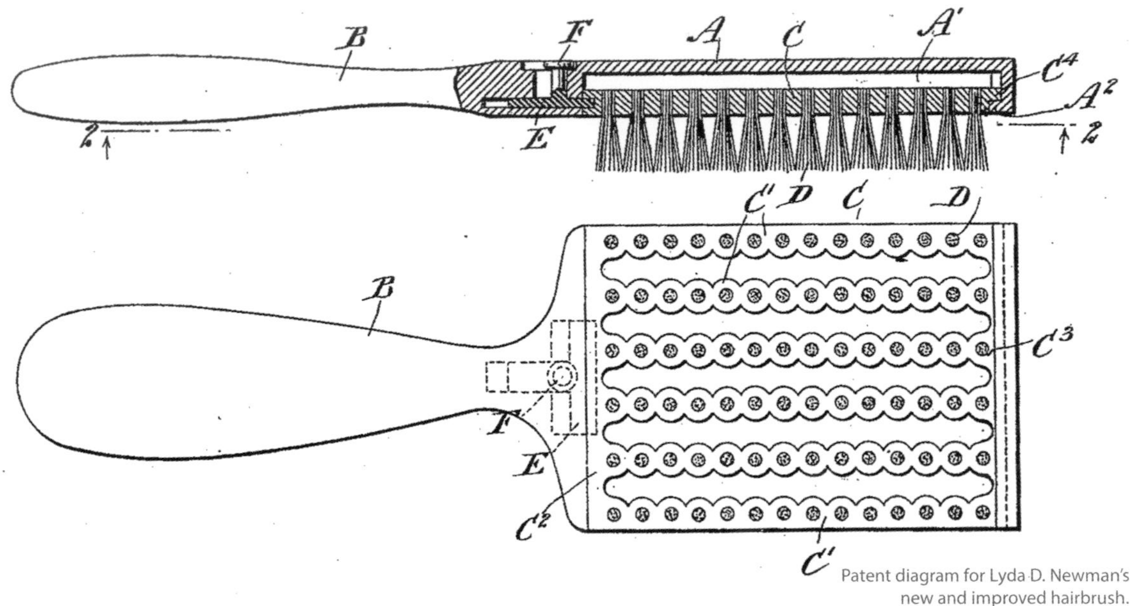

Patent diagram for Lyda D. Newman's new and improved hairbrush.

African American women were important in the development of haircare products in the nineteenth century, and two of the best examples are Madam C.J. Walker and Marjorie Joyner, but Lyda D. Newman also played a significant part. A hairdresser by trade, she obviously did not invent the hairbrush; it had been around for a long time. What she did give the world, however, was a new and improved type of hairbrush that could be easily cleaned.

We know little about Newman's life but can safely presume that as a black child at that time, she would have had little more than a few years of elementary education, at most. At the time she was granted her patent, however, we do know that she was residing in Manhattan. She also fought for women's right to vote, working with well-known women's suffrage activists.

The hairbrush she patented was simple and durable, worked very well in its function of brushing the hair but was also easy to clean. The impurities that come from the scalp or hair as it is brushed would pass through openings or slots in the brush and into a recess at the back. These were removed by disconnecting the holder from the brush and simply blowing these impurities out of the back's open sides. It was far from a harebrained idea!

THE 1876 PHILADELPHIA CENTENNIAL EXPOSITION

The Centennial Exposition, held in Philadelphia from May to November 1876 was America's first official World's Fair, staged in celebration of the centenary of the signing in Philadelphia of the Declaration of Independence. The fair was visited by almost 10 million people and thirty-seven countries were represented.

After much aggressive lobbying by early feminists and women's suffrage movements, it was agreed that women would be represented by a separate Women's Pavilion devoted entirely to the artistic and industrial endeavors of women.

More than eighty inventions patented by women were displayed, including a reliance stove, a hand attachment for a sewing machine, a dish-washer, a heating iron with removable handle, a frame for stretching and drying lace curtains, and a stocking and glove darner.

ALICE PARKER

CHANGING THE WAY BUILDINGS ARE HEATED
INVENTION: A GAS HEATING FURNACE

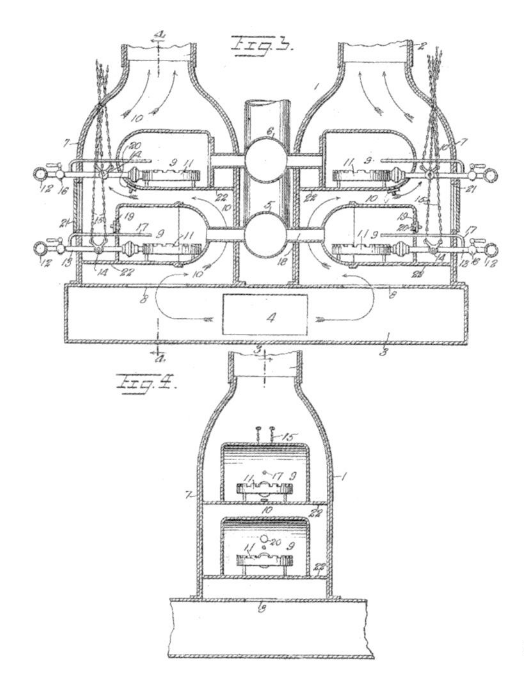

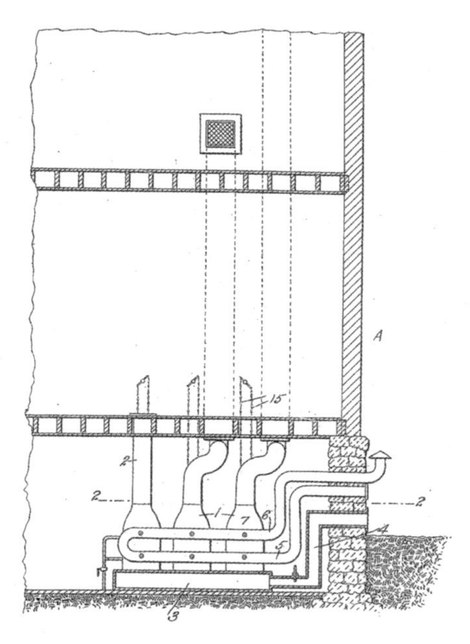

Patent diagram for an improved heating furnace for houses or buildings to supply independently controlled heating to separate rooms or floors.

We know next to nothing about Alice Parker, apart from the fact that she was an African American woman from Morristown, New Jersey, who studied at Howard University in a time when few African Americans went to university. In 1919 she came up with an idea that changed the way buildings were heated.

Central heating was by no means a novel concept in the early years of the twentieth century. Indeed, the Ancient Romans used a system of central heating, having slaves fuel a furnace, the heat of which was circulated under the floors and through the walls of Roman villas. In the eighteenth and nineteenth centuries, systems were designed to create warmth and, in fact, by 1888 the United States Patent Office had issued more than 4,000 patents for heating stoves and furnaces.

Parker's system was an early version of the modern home heating system. It gave rise to the thermostat and was a precursor of modern forced air furnaces. It was also labor saving as the most common method of heating a home was by burning wood that had to be chopped. Parker's system was unique, however, a natural gas-fueled "new and improved heating furnace" with the warm air being distributed through ducts into the rooms of the house. It was the first time anyone had considered using natural gas for domestic heating.

We are unaware if anyone actually purchased Parker's system and installed it—it was notoriously difficult for African Americans to have their products taken seriously, let alone an African American *woman*—but it is now viewed as an important precursor to the modern central heating system.

HEATING THE HOME

It was the ancient Greeks who first used central heating. At the temple of Ephesus, a fire provided the heat which was circulated via flues that were sunk into the ground. The Romans, too, were no strangers to the circulation of heat from a central source. Villas have been found in which warm air created by furnaces was circulated under floors and through ducts in the buildings' walls. This system is known as hypocaust, deriving from the Ancient Greek hypo meaning "under" and caust meaning "burnt."

FLORENCE PARPART

A WOMAN OF MANY PARTS

INVENTION: THE STREET-SWEEPER AND THE REFRIGERATOR

A salesman demonstrating an electric refrigerator to a potential buyer in the 1930s.

Florence Parpart is yet another inventor about whom we know very little. Some patent applications and census records are about all there is. She was born in Hoboken, New Jersey, and was, for the majority of her life listed in the records as a housewife, living in New York City and Pittsburgh.

Parpart's first patent was in 1900 for an improved street-sweeper which had been invented by Eureka (the perfect name for an inventor) Frazer Brown in 1879. Apparently, she had become annoyed when her dress was splashed with mud from a street-sweeping machine and she resolved to make improvements to it. Her fiancé at the time, Hiram D. Layman, was a skilled electrician and it is likely that he helped her to design her prototype. Within a couple of years, she had contracts to supply her improved street-sweeper to a number of American cities.

Florence Parpart's second patent, and the one for which she is best known, was in 1914 for the modern refrigerator. Up to that point, an icebox had been used to keep foodstuffs fresh but Parpart's new fridge used electricity. By this time she had married Hiram Layman and again his skills came to the fore. They jointly patented the Refrigerator via US Patent 1,090,925. A skilful marketeer and salesperson, she demonstrated and sold her refrigerator at trade shows and advertised it, turning it into a commercial success.

By the 1950s refrigerators had got bigger and better!

FRANCES PAULET

CREATING BLUE-VEINED CHEESE

INVENTION: STILTON CHEESE

Stilton is an English Blue cheese, known for its characteristic strong smell and taste.

Frances Paulet (sometimes "Pawlett"), a skilled cheese-maker of Wymondham in Leicestershire, England, is said to have created the blue-veined cheese we know as Stilton in the 1720s. Stilton is made golden with double cream, the cream of one milking being added to the milk of the next. There is a great deal of debate about the cheese's origins, but the fourteenth edition of the *Encyclopedia Britannica* says:

> *Mrs. Paulet of Wymondham in the Melton district of Leicestershire, is said to have been the first maker of Stilton cheese. She supplied them to Cooper Thornhill, who kept the Bell Inn at Stilton in Huntingdonshire on the great north road from London to Edinburgh, and they became famous among his customers, and throughout England. The manufacture of Stilton cheeses became an industry of the district. Mrs. Paulet was still living in 1780.*

Stilton cheese can now only be made in three English counties—Derbyshire, Leicestershire, and Nottinghamshire, and at present just six dairies are licensed to make it. Curiously enough, it cannot be made in the village that gave it its name. The village of Stilton is in the county of Huntingdonshire.

LILY PAVEY

A TYPEWRITER THAT WRITES SHEET MUSIC
INVENTION: THE MUSIGRAPH OR MUSIKRITER

Musician Lily Pavey writing sheet music on an adapted Remington typewriter, 1963.

Many had tried and failed to invent a music typewriter, one that when you tap the key, sounds a note which it prints in the correct place on a sheet of music paper. With her Musigraph or Musikriter, patented in 1961, however, Britain's Lily Pavey succeeded in doing just that. Not that she had a background that was conducive to the creation of something as complex as her machine. She was born in a circus and did not have a great deal of education but she was possessed of an intuitive mathematical ability, as well as, apparently, a dream "Voice" that guided her through her work.

The Voice, she claimed, instructed her to invent her music typewriter although she did first conceive of it while riding in an elevator. She worked at night for thirteen years trying to convert a normal typewriter, working out how she could provide it with vertical elevation without having to move the paper. There was also the small matter of creating 8,000 different combinations with just 46 keys.

Pavey was a humble office worker and, therefore, had to be supported in her work by backers such as the Gulbenkian Foundation but soon her machine was receiving plaudits. Newspapers talked of how "Little Miss Pavey has done something which has always baffled the world's engineers and musicians." She also picked up the gold medal at the Brussels Inventors' Exhibition in 1963.

A month later, she created a company at Dorking, Surrey, to manufacture her machines, selling them for £100 each. Offers from companies wanting to be involved with her came from as far afield as France and Japan, but eventually she went with the Imperial Typewriter Company who sold her machines for £260 each. Unfortunately, sales were far from impressive.

She soldiered on and even devised a new machine, the Electronome which allowed a composer to hum into a microphone, the note being printed out by a typewriter. She also developed a typewriter named the Spherigraph that could add words to the music as well as do ballet notation and type complex mathematics and chemistry symbols.

ANNA PAVLOVA

GET MY SWAN COSTUME READY

INVENTION: THE MODERN BALLET SHOE

You might think that she invented the famous meringue dessert, the deliciously sugary pavlova, but, that seems to have been invented by a chef in Wellington, New Zealand, when the ballet dancer Anna Pavlova performed there in 1926 during her world tour.

Anna Pavlova (1881 – 1931) was one of the most celebrated ballerinas in dance history. She was inspired to become a ballerina from a performance of *The Sleeping Beauty* in St. Petersburg, Russia, where she was born. At the age of 10, she was accepted by the Imperial Ballet School, graduating at 18. She was chosen to enter the Imperial Russian Ballet where in 1905 her most famous role, *The Dying Swan*, was created for her by the groundbreaking Russian choreographer and dancer, Michel Fokine.

Anna Pavlova developed the modern ballet shoe to protect her feet from injury while dancing.

In 1909, she performed in Paris with the troupe the Ballets Russes assembled by Sergei Diaghilev but was unhappy with Igor Stravinsky's avant-garde music for *The Firebird* and refused to dance it. In 1913, she bought Ivy House in London which was her home from then on. A year later she founded her own company and began to tour the world.

Anna Pavlova can be credited with the creation of the modern ballet shoe. Ballet had first been danced by women in 1682 and the standard ballet shoes back then had heels. The non-heeled shoe was first worn by Marie Camargo of the Paris Opéra Ballet in the middle of the eighteenth century and by the end of the century, ballet shoes did not have heels.

Pavlova had very rigid feet, high, arched insteps, and slender, tapered feet which made her vulnerable to injury when she was dancing *en pointe*, when the body weight is supported on the tips of the toes. She added a piece of toughened leather on the soles for support and flattened and hardened the area around her toes to form a kind of box. With her modifications to her shoes making *pointe* work less painful and easier for curved feet, her approach soon became standard.

In 1931, while performing in The Hague, Belgium, Pavlova became ill with pneumonia. She was told by doctors that she required an operation but was warned that afterward she would never be able to dance again. She refused to have the operation, insisting that she would rather die than never dance again. Three weeks before her fiftieth birthday in 1931, she died of pleurisy, her last words were reported to be: "Get my *Swan* costume ready."

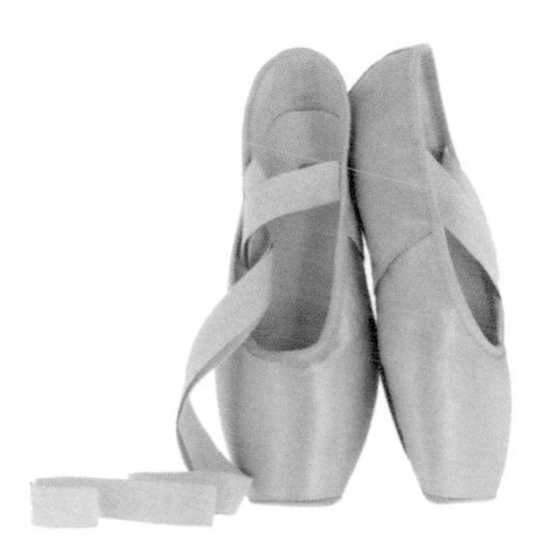

Modern-day ballet shoes.

MARY ENGLE PENNINGTON

THE ICE WOMAN

INVENTION: COLD STORAGE

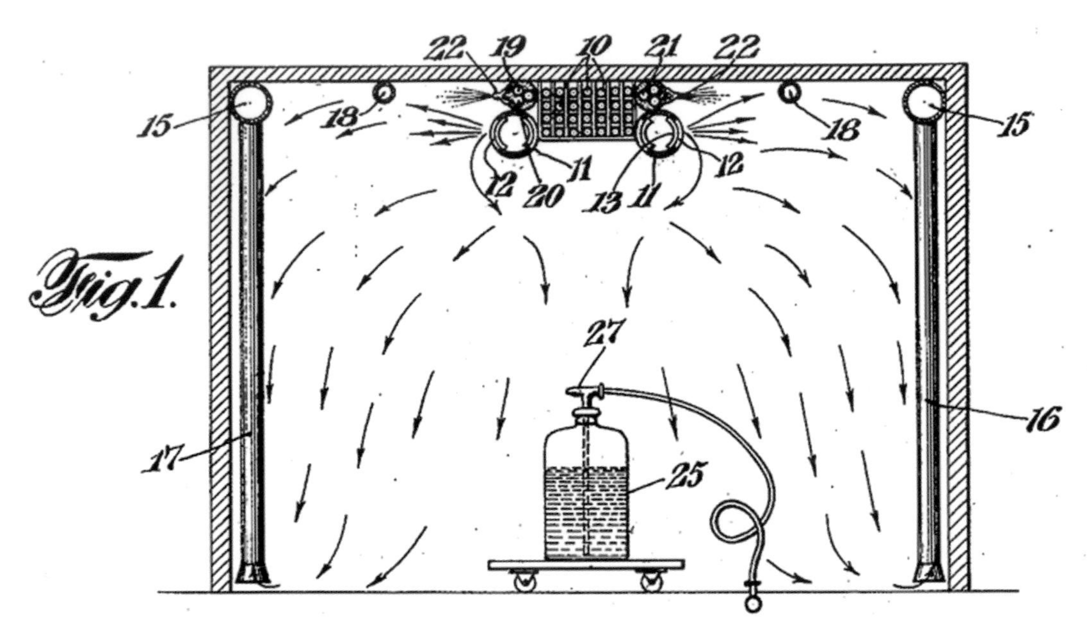

Patent diagram for air conditioning systems for cold storage rooms.

Mary Pennington's interest lay not in the refrigerator itself, but rather in the theory and the methods that could be used in a national system of frozen food distribution. That did not mean that she did not have specific inventions to her name, however. Dr. Pennington improved both residential and commercial refrigerators as well as cold storage units.

Mary Pennington was born into a Quaker family in 1872 in Nashville, Tennessee, and from an early age was fascinated by chemistry. She graduated from the University of Pennsylvania in 1892 with a science degree, but, as was common in those times, the university did not grant degrees to women. Therefore, she was awarded with a proficiency certificate instead. She gained a PhD from the same institution in 1895 and was a fellow in chemistry at Yale between 1897 and 1899. In the coming years, she worked at the Women's Medical College of Pennsylvania, at the department of hygiene at the University of Pennsylvania, and at the Philadelphia Bureau of Heath. In 1905, she was given a position as a bacteriological chemist at the US Department of Agriculture and in 1907 was appointed head of the newly created Food Research Laboratory.

At the Food Research Laboratory, she worked on the design of refrigerated boxcars and this started her interest in the transportation and storage of perishable food. By 1913, she was experimenting with the freezing of sweet-corn kernels in Minnesota, studies during which she solved the problem of controlling humidity during freezing. Without that, foods were likely to dry out but if they were too moist, mold would set in.

In the fish industry, Dr. Pennington invented a process for scaling, skinning, quick-freezing, and dry-packing fish-fillets as soon as a catch was landed. As with many of her inventions, this process became commonplace in the industry.

Not for nothing was Mary Pennington described in a 1941 *New Yorker* profile as the "Ice Woman." She served nothing but frozen food at her own table at home, boasting of serving food that came from every part of the United States, all brought to her table because of refrigeration.

JUDY W. REED

FIRST PATENT FOR AN AFRICAN AMERICAN WOMAN

INVENTION: A DOUGH KNEADER AND ROLLER

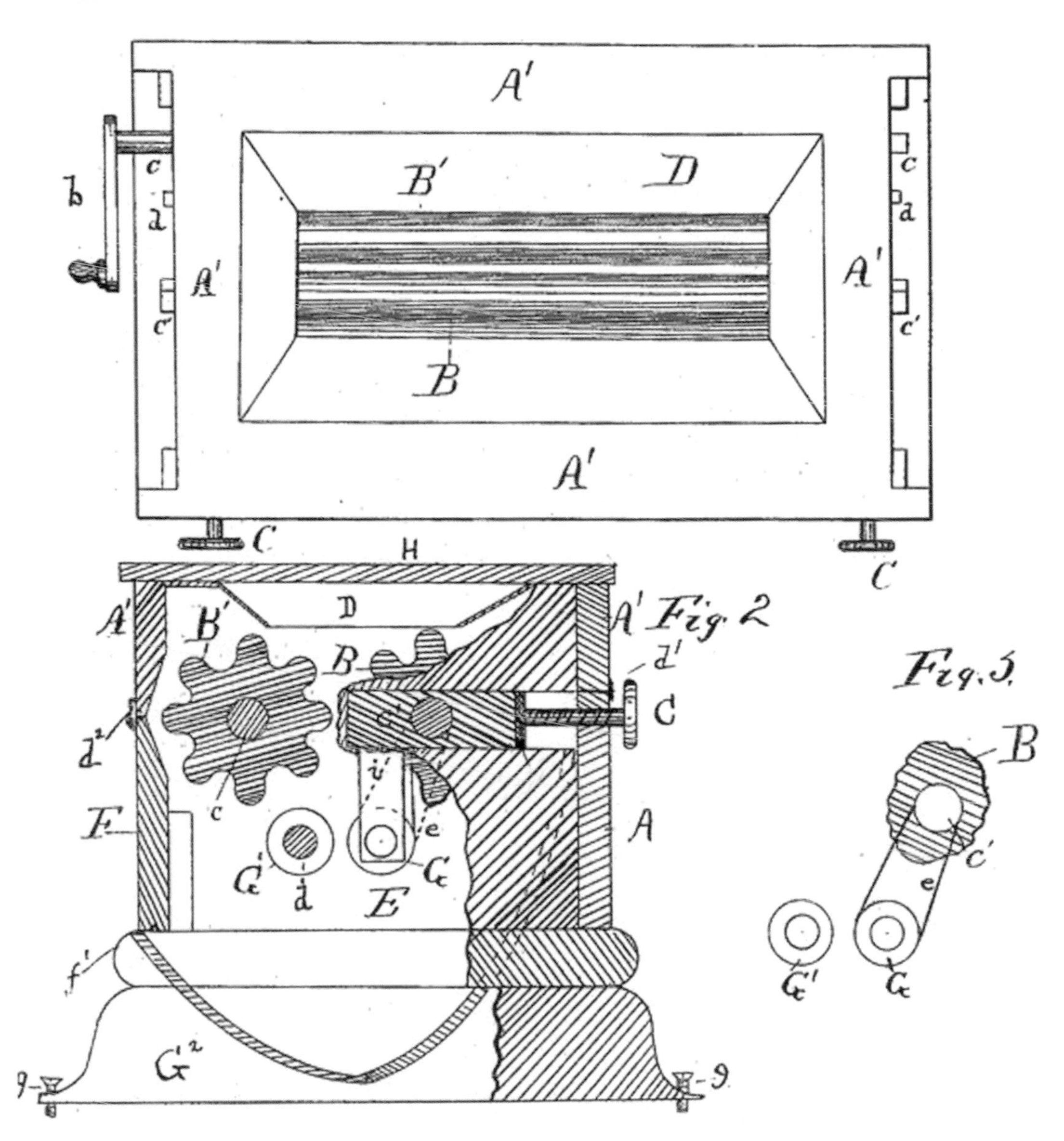

Judy W. Reed's patent diagram for new and useful improvements in dough kneaders and rollers.

Judy Reed (*c.* 1826 – ?) is yet another African American woman about whom we know very little, although she is thought by some to be the first African American woman to be awarded a patent. She applied for one in January 1884 for an improved design for existing dough kneaders. Her device mixed the dough much more evenly as it was processed through two rollers that had corrugated slats carved into them that would act as the kneading implements. The dough was then delivered into a covered receptacle which would serve to keep it clean and protected from the elements. Uneducated and illiterate, Reed signed patent number 305,474, granted in the name of J.W. Reed on September 23, 1884, with an *"X."*

ANGELA RUIZ ROBLES

PIONEER OF THE ELECTRONIC BOOK. INVENTION: A BOOK READER

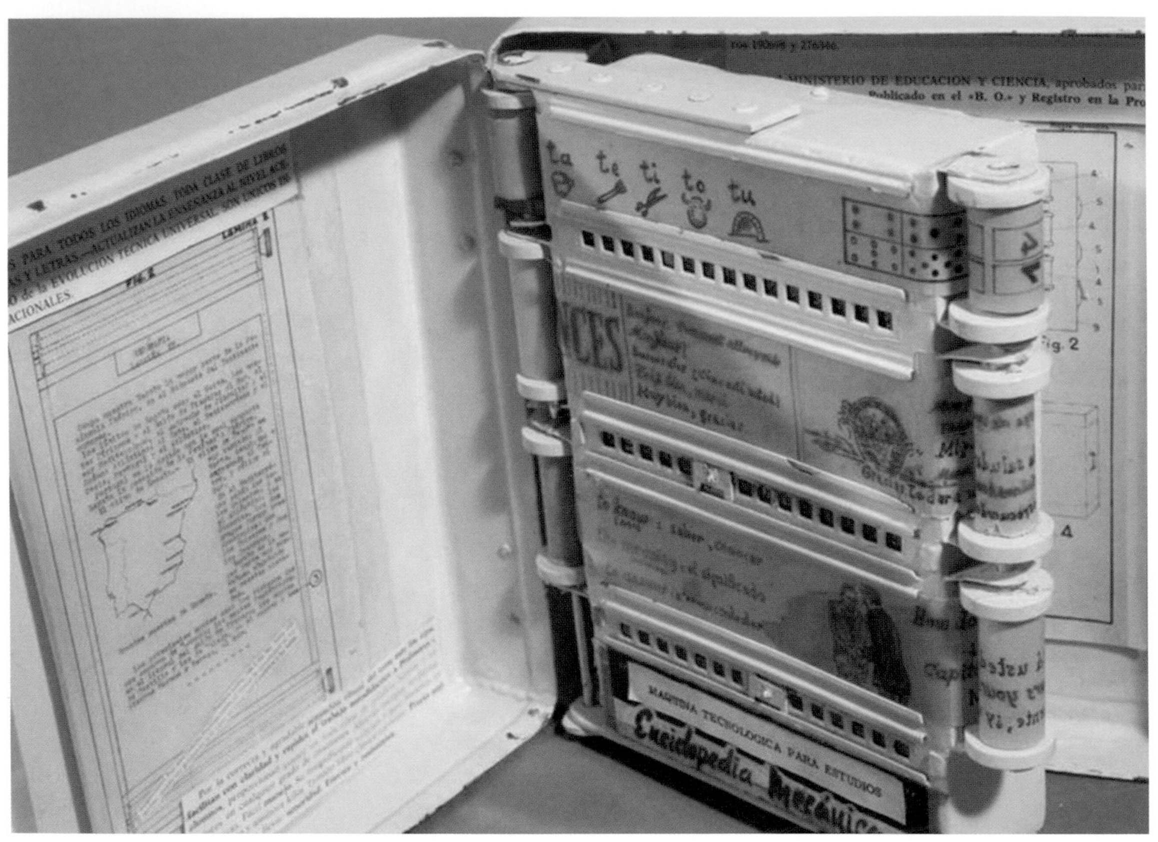

The "Mechanical Encyclopedia" was never put into production but the prototype in the photo is kept in A Coruña National Science and Technology Museum, Spain.

We think nothing nowadays of grabbing our cell phone, tablet or e-reading device and devouring a few pages of a new bestseller, presented digitally without pages or covers. It may surprise many to know, however, that the first book reader was invented as far back as 1949 by a Spanish woman, Angela Ruiz Robles (1895 – 1975).

Robles was a Spanish teacher in Franco's Spain but was concerned at the weight of the books that her students had to haul around with them. So she created the world's first automated reader, calling it *la Enciclopedia Mecánica*—"the Mechanical Encyclopedia."

The reader operated on compressed air, and text and graphics were on spools that users could load onto rotating spindles. Everything came in a durable metal case. She said: "It has some coils [or spools] where you place the books that you want to learn in whatever language. By a movement of the same [coils or spools] it passes over all the topics, making it stop where you like it to."

Sadly, although she tried until she passed away in 1975, she was never granted a patent for her device as it was believed to be impractical to attempt to build it. Later in her life when technology had improved, she was unable to raise the funds necessary for the revival of her project.

ERNESTINE ROSE

IN SUPPORT OF WOMEN'S RIGHTS

INVENTION: A ROOM DEODORIZER

A selection of nineteenth-century room deodorizing booklets made by Papier d'Armenie Company, Paris, France.

Ernestine Rose, born Ernestine Louise Potowski in Poland in 1810, was a reformer and suffragist, a champion of women's rights in the nineteenth century, and a supporter of temperance and the abolition of slavery. Her father was a rabbi who had married the daughter of a wealthy businessman which allowed Ernestine the opportunity (rare among young Polish Jewish girls) to be educated in the scriptures. She rejected such notions as women being inferior to men, however, by her early teens she had also rejected the idea of the existence of God.

Her struggles against the customs and religious norms began when she was 16. Her father had chosen a man for her to marry but, horrified at the idea of being married to someone she did not love and barely knew, she took her case to the civil courts and won. Leaving Poland, she lived in Berlin, Germany, supporting herself with sales of her invention of a room deodorizer. It was a simple but effective invention using perfumed papers in a room to kill all the unpleasant odors inside and the ones coming from outside. Commercially successful, it enabled her to fund further travels.

She voyaged around Europe before heading for England but, having lost everything in a shipwreck, she was forced to support herself by teaching and again selling her room deodorizers. While in England, she met the social reformer Robert Owen and together they founded an organization, the Association of All Classes of All Nations. As the name suggests, the group advocated equal rights for all, no matter their ethnicity, sex, or class. Ernestine also married a jeweler named William Rose and the two decided to emigrate to the United States.

That same year—1836—she petitioned the New York State Legislature in an effort to obtain a married woman's property act. She failed to obtain the requisite support, but it was the first petition brought before the legislature in support of women's rights. Before long, she was touring America, speaking out against slavery and advocating women's rights, a dangerous occupation at the time. In 1854, she was voted president of the National Women's Rights Convention, a controversial election because of her avowed atheism.

She continued to campaign for women's rights until moving back to England with her husband in 1869, with her health deteriorating. She started campaigning again in 1873, throwing her weight behind women's suffrage, and continued to do so until her death in 1892, at age 82. Her life was a shining beacon for anyone who cares about freedom and human rights, but it might not have turned out the way it did if she had not come up with a brilliantly simple idea for a room deodorizer while living in Berlin.

LADY ANNE SAVILE

WHAT GOES UP MUST COME DOWN

INVENTION: AN ANTI-SEASICKNESS BED

The English aristocrat and socialite, Lady Anne Savile (1864 – 1927), daughter of the Earl of Mexborough, had a lifelong passion for aviation when flying was still new and extremely dangerous. In the end she flew to her death as she attempted a perilous transatlantic crossing.

Born in 1864, at the age of 33 she married Prince Ludwig of Löwenstein-Wertheim-Freudenberg, becoming a citizen of the German Empire. A year later, in March 1899, her husband disappeared while fighting in the Philippines in the Spanish-American War. Princess Anne never remarried but held onto her title, even after the start of World War I when German titles were not popular in Britain. However during her participation as a passenger in aviation events, she usually flew under her maiden name, Lady Anne Savile.

Before the war, Anne traveled a number of times to the United States. In 1913, she boarded the White Star liner, SS *Majestic* in Southampton, England, accompanied by her own invention, an "automatic balancing bed" that she declared was guaranteed to prevent seasickness. Sadly, not much else is known about it or whether it performed as she promised when she sailed to New York.

In 1927 Anne died while attempting to be the first woman to fly the east to west crossing of the North Atlantic Ocean. She was a passenger in the *Saint Raphael,* a monoplane piloted by World War I flying ace Captain Leslie Hamilton who was nicknamed the "Flying Gypsy." Anne, Captain Hamilton, and a third crew member Colonel Frederick F. Minchin disappeared in flight. Their aircraft was last seen about 800 miles west of Galway, Ireland, flying into a fog bank. It was assumed they had all perished after plunging into the ocean.

THE MYSTERY OF AMELIA EARHART

Following in the daredevil footsteps of Lady Anne Savile, Amelia Earhart (1897 – 1939) was an equally intrepid American aviator who was also lost at sea. Amelia set many flying records and championed the advancement of women in aviation. Her notable flights include being the first woman to fly across the Atlantic Ocean in 1928 as well as the first person ever to fly solo from Hawaii to the US mainland. During a flight around the world, Amelia disappeared somewhere over the Pacific in July 1937. Neither her plane nor her body were ever found and she was officially declared dead in 1939. Her disappearance remains one of the great unsolved mysteries of the twentieth century.

Amelia Earhart was a widely known international celebrity during her lifetime and her disappearance has sustained her lasting fame in popular culture. Hundreds of articles and books have been written about her life, and she is generally regarded as a feminist icon.

Amelia Earhart after landing at Culmore, Northern Ireland, May 21, 1932.

MARGARET CLAIRE SHEPHARD

WORKING AT THE FRONTIERS OF KNOWLEDGE
INVENTION: AGRICULTURAL FUNGICIDES

Farm tractor spraying a soybean field with herbicides, pesticides, and fungicides.

Brought up in Maidenhead, England, in an environment where "defeat was not encouraged," Dr. Claire Shephard (born 1931) is the holder of at least twenty patents with various co-inventors for fungicides, pesticides, and agents that control plant growth. Her expertise developed from the study of plant pathology at the Imperial College of Science and Technology, after she had earned a BSc in botany, with honors, from University College London in 1952. She completed a PhD in plant pathology in 1957.

Hoping to find chemical ways of helping food production in a world that was increasingly desperate for food, she went to work in the research department of Imperial Chemical Industries (ICI) in 1954 and remained there working with a team inventing new fungicides for use in agriculture. The importance of such work can be demonstrated by the fact that fungal diseases are responsible for the waste of billions of dollars of crops around the world every year. Her work is pioneering, often dealing with new chemicals whose properties are unknown, as she has said: "This has entailed working at the frontiers of knowledge."

Dr. Shephard has invented and patented new chemicals and has used untested compounds to make innovative new fungicides that are practical and effective. She has had to design and create techniques for testing her compounds on small potted plants instead of on large fields of plants, and she has had to create facilities which provide the appropriate environment for her trials as well as new techniques for handling plants and plant diseases in the glass house. She is part of the team that created dimethirimol, ethirimol, and bupirimate which, although not the first systemic fungicides, were the first commercially available products that could be used on both soil and seed to deal with diseases of foliage.

PATSY SHERMAN

GIRLS SHOULD FOLLOW THEIR DREAMS
INVENTION: SCOTCHGARD STAIN REPELLENT

One of very few women chemists in the 1950s, Patsy Sherman co-created Scotchgard, a stain repellent and remover to keep fabrics looking fresh.

How often are brilliant inventions the result of mistakes by scientists when trying to do something else? Scotchgard, the durable stain- and water-repellent is a case in point. Patsy Sherman (1930 – 2008), the inventor of Scotchgard, was a chemist working for the multinational company 3M, who had been delegated to develop a rubber material to withstand contact with jet aircraft fuels. One day, as she was experimenting in her laboratory, an assistant spilled some chemicals she had been working with onto her white tennis shoes. She immediately tried to clean off the marks but nothing she tried was effective.

She continued to wear the shoes and from then on, as the rest of the shoe became dirty through wear, that spot remained sparkling white. Sherman realized that it was the spillage that was keeping the shoe good as new and that there was perhaps a commercial application for this. For several years, she worked with fellow scientist, Samuel Smith, developing a fluorochemical polymer that was able to repel oil and water from fabrics.

Their first patent was granted in April 1971 for the "invention of block and graft copolymers containing water-solvatable polar groups and fluoroaliphatic groups." They called their product Scotchgard and it became the most widely used stain-repellent and soil-removal product in America. Products proliferated and eventually, Sherman held sixteen US patents, thirteen of which were shared with Samuel Smith.

Patsy Sherman became a chemist through sheer determination. When she was 17, she took a high school aptitude test that told her that she was best suited to the role of housewife. Irritated by this, she took the boys' test and its results said conversely that she should look to science or dentistry for her future career. Having graduated with degrees in chemistry and mathematics from Gustavus Adolphus College in St. Peter, Minnesota, she began working at 3M in 1952.

Patsy was a scientist at a time when women in science were a rarity. In fact, when she was developing Scotchgard in the 1950s she was not even allowed inside the textile mill during product testing because women were banned. She once said, "girls should follow their dreams. They can do anything anybody else can do," and she was living proof of that.

BEATRICE SHILLING

MISS SHILLING'S ORIFICE
INVENTION: THE RAE RESTRICTOR

Rolls-Royce Merlin engines powered the Spitfires and Hurricanes of World War II.

During World War II a disastrous deficiency developed in the Merlin engines of the British Royal Air Force's iconic Hurricane and Spitfire fighter planes. Whenever the planes dived, their carburetors flooded with fuel, causing the engine to cut out and leaving the pilot at the mercy of the enemy. The problem was solved by a woman, Beatrice "Tilly" Shilling, an engineering graduate from Manchester University. It can easily be argued that her invention helped to win the war.

Tilly Shilling (1909 – 1990) was born in Waterlooville, Hampshire, England, the daughter of a butcher. From an early age, she had resolved to be an engineer. At age 14, she bought a motorcycle just to learn how to take it apart and put it back together. Leaving school, she spent three years working for an engineering company before studying electrical engineering at Manchester University.

She graduated in 1932 but continued studying for a Master of Science degree in mechanical engineering. She found work as a research assistant at Birmingham University before becoming a scientific officer at the Royal Aircraft Establishment (RAE) in Farnborough, Hampshire, where she remained until her retirement in 1969.

Tilly Shilling fixed the engine problem with a simple device called the RAE Restrictor. It was a brass thimble with a hole in the middle (later simplified to a flat washer), which could be fitted into the engine's carburetor without taking the aircraft out of service. This allowed just the correct amount of fuel required for maximum engine power and maneuverability during dogfights.

By March 1941, Tilly and her team had installed her device in every Merlin engine in RAF Fighter Command. With typical humor, the restrictor was nicknamed "Miss Shilling's orifice" by the battle-hardened pilots. Although it was merely a stop-gap solution, it was simple and elegant and dealt with the problem until 1943 when pressure carburetors were introduced.

Following the war, Miss Shilling worked on Britain's medium-range ballistic Blue Streak missile. She was a staunch advocate of equality for women, refuting vehemently that women were inferior to men. She also loved speed. She raced motorbikes in the 1930s and, with her husband George, raced cars after 1945. She died at age 81 in 1990.

AMY B. SMITH

REACHING OUT ACROSS THE WORLD
INVENTION: THE SCREENLESS HAMMER MILL AND PHASE-CHANGE INCUBATOR

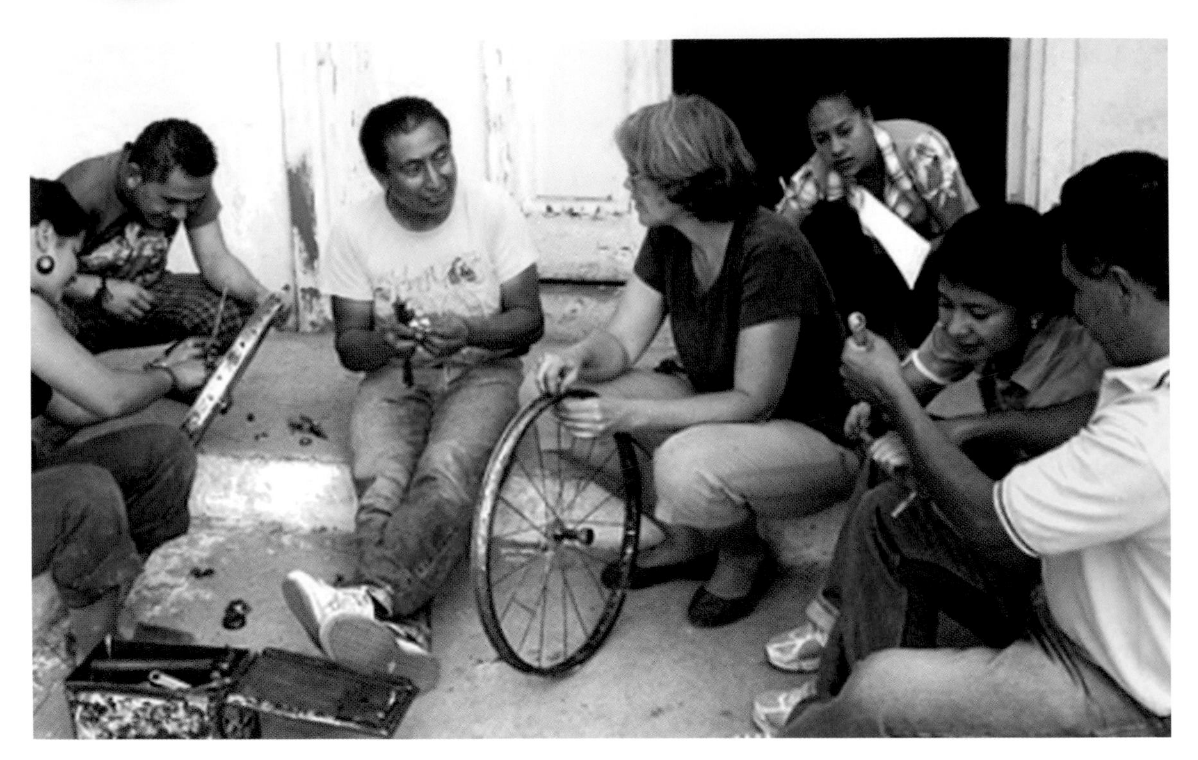

Amy Smith helping with a bicycle repair group in El Salvador, 2014, as part of her D-Lab outreach mission to people all round the world to develop self-help practical solutions and initiatives.

Amy B. Smith (born 1962) reaches out to help people across the world who are worse off than she is, with technologies that improve their quality of life as well as their health. She works in engineering design and appropriate technology for developing countries at the Massachusetts Institute of Technology, aiming to teach students about the problems of the Third World so that they can potentially help in the future. She also works to encourage collaboration between researchers around the world in order to come up with medical technologies.

Among Smith's inventions are the screenless hammer mill and the phase-change incubator. The former is a motorized hammer mill that is used to pound grain into flour. Normally a screen is used in this but these are subject to breakage. Smith's version uses the flow of air to segregate small particles and large ones. Her invention makes the device 25 percent cheaper and more reliable.

The phase-change incubator is a cheap incubator that can test for microorganisms in drinking water, using small balls containing a chemical compound that is heated and that stays at 37°C (99°F) for 24 hours. Cultures can be tested without having to be sent away to a laboratory. Needless to say, it is of particular value to isolated communities.

Smith, who was born in Massachusetts, is involved in a number of initiatives that aim to develop technologies using a network of creative people internationally.

HARRIET WILLIAMS RUSSELL STRONG

FASCINATED BY WATER
INVENTION: WATER STORAGE AND FLOOD CONTROL

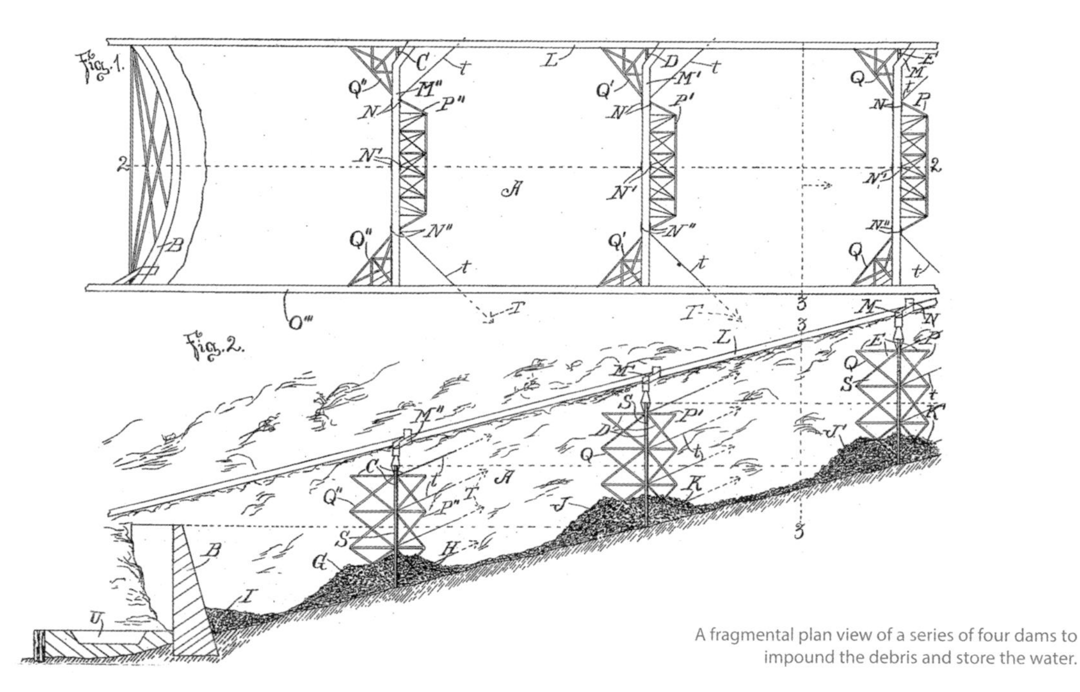

A fragmental plan view of a series of four dams to impound the debris and store the water.

After being widowed in 1883, Harriet Williams Russell Strong (1844 – 1926) was left property in the San Gabriel Valley in southern California and to support herself and her four daughters, she had to manage and develop the land. Ranchito del Fuerte, as it was called, was mainly planted with profitable walnut, pomegranate, and orange trees, and Harriet also cultivated fast-growing pampas grass whose plumes she sold to the millinery trade.

She purchased 1,000 acres of land nearby to utilize water she had discovered after drilling artesian wells. After installing a pumping plant, she established the Paso de Bartolo Water Company. She sold the property several years later and with the profit she made, wiped out all her family's debt.

Strong had by this time become fascinated by water and the problems associated with it. Of particular interest were the control of flood waters and the efficient storage of it in dams and reservoirs. Her first patent, granted in December 1887 was for water storage. The clever idea she presented was for a series of dams, situated one behind the other that would be built in a valley, canyon or watercourse. They would be constructed so that each dam, when filled with water, would act as a support for the one in front of it.

She appeared before the congressional committee on water power in 1918 to urge the government of the day to store the flood waters of the Colorado River by building a series of dams, using her method, in the Grand Canyon. It did not happen, but her work undoubtedly led to the construction of the Hoover Dam and the All-American Canal in south-eastern California.

In 1894, she obtained a second patent "for a convenient, cheap, and effectual impounding of debris from hydraulic mines, settling the water, and storing the same so as to allow it to be used for irrigation or other purposes." These were not her first patents. She had earlier patented a design she had come up with for a device to raise and lower an upper window sash.

This astute businesswoman, brilliant inventor, social activist, conservationist, and leading light of the early women's movement died in a car crash in 1926, at age 82.

BRIDGET ELIZABETH TALBOT

SAVING THOUSANDS OF LIVES

INVENTION: A WATERPROOF ELECTRIC TORCH FOR LIFEBELTS

The USCGC *Escanaba* rescuing the survivors of SS *Dorchester* in the predawn darkness of February 3, 1943. The seamen did not have waterproof torches but the rescue was marked by the *Escanaba's* historic first use of rescue swimmers in survival suits.

Bridget Elizabeth Talbot (1885 – 1971) was an extraordinary British politician and campaigner who crammed a remarkable amount into her 86 years. The daughter of the Honorable Alfred Chetwynd Talbot and Emily Louisa Augusta, she worked with the Red Cross and refugee charities across Europe during World War I and continued her tireless charity work after the war. Italy awarded her the Italian Medal for Valor for her work with the Anglo-Italian Red Cross on the Italian-Austrian front and she received an OBE from King George V in 1920.

Between 1920 and 1922 she was in Turkey, trying to deal with Russian refugees. After that she ran a cooperative farm colony in Asia Minor. In 1931, she joined the Labour Party, working with then Prime Minister Ramsey Macdonald on the election campaign of that year and she became a member of the National Labour Council.

In 1932, she was in Russia with the British philanthropist and humanitarian worker, Lady Muriel Paget establishing a convalescent home for "Displaced British Subjects," British residents in the Soviet Union who had been unable to leave the country after the 1917 October Revolution because they were unwilling, too old, too infirm, or too poor.

One of the causes to which Talbot devoted her attention was the working conditions of British merchant seamen. In 1937, she even gained first-hand experience by serving on a sailing ship voyaging to Finland. Two years later she launched an enquiry on behalf of the National Labour Party into the Merchant Navy.

This interest in seamen and the dangers that they encountered may have led to an invention by her that saved hundreds and possibly thousands of lives during World War II. It was a waterproof electric torch for lifebelts to help sailors be located when they had been forced to abandon ship or had been thrown into the water during enemy action. Not only did she invent the torch, however, in her very determined way, using her network of political and social connections, she succeeded in having these made compulsory for all Merchant Navy, Royal Navy, and Royal Air Force personnel in Britain.

MÁRIA TELKES

THE SUN QUEEN
INVENTION: SOLAR ENERGY CONVERSION

Born in Budapest, Hungary, biophysicist Mária Telkes (1900 – 95) was far ahead of her time. A pioneer in the use of solar energy to heat our homes and our cars, not for nothing was she dubbed the "Sun Queen." In fact so far ahead of her time was she that it is only in recent decades that her ideas have been put into practice, and her huge contribution to science has been recognized.

After graduating from Budapest University with a doctorate, she was offered a position as a biophysicist at the Cleveland Clinic Foundation, Ohio, USA. In 1937, she was granted American citizenship and moved to a new position as a research engineer at Westinghouse Electric, the American manufacturing company.

In 1939, she began work as a research associate in metallurgy at the Massachusetts Institute of Technology and her attention began to turn very much toward solar energy. She and her team were part of the Solar Energy Conversion Project which built the Dover Sun House, in Dover, Massachusetts, completed in 1948. It was heated purely by solar energy using a thermoelectric power generator invented by Telkes.

During World War II, she worked as a civilian advisor to the Office of Scientific Research and Development, creating a solar still for life rafts to distill sea water, using the heat of the sun to make it suitable for drinking. Her portable desalination unit was widely used by the US military during the war, a version of it is still in use today.

In 1953 Dr. Telkes moved to New York University where she established a solar energy lab. Five years later, she became Director of Research for Solar Energy at the Curtiss-Wright Company developing solar dryers and water heaters. She also worked on thermoelectric generators for spaceships. In the 1960s and 1970s she was granted several patents for methods of storing solar energy.

The sun finally set on the "Sun Queen" in 1995 and she passed away at the age of 95. The potential of her vision of a solar-powered future is only now being realized by others. She was inducted into the National Inventors Hall of Fame in 2012.

The solar-powered house of the future, illustrated on the cover of the March 1949 edition of *Popular Science*.

TEMPLE GRANDIN

EMBRACING YOUR INNER SELF

INVENTION: THE HUG BOX

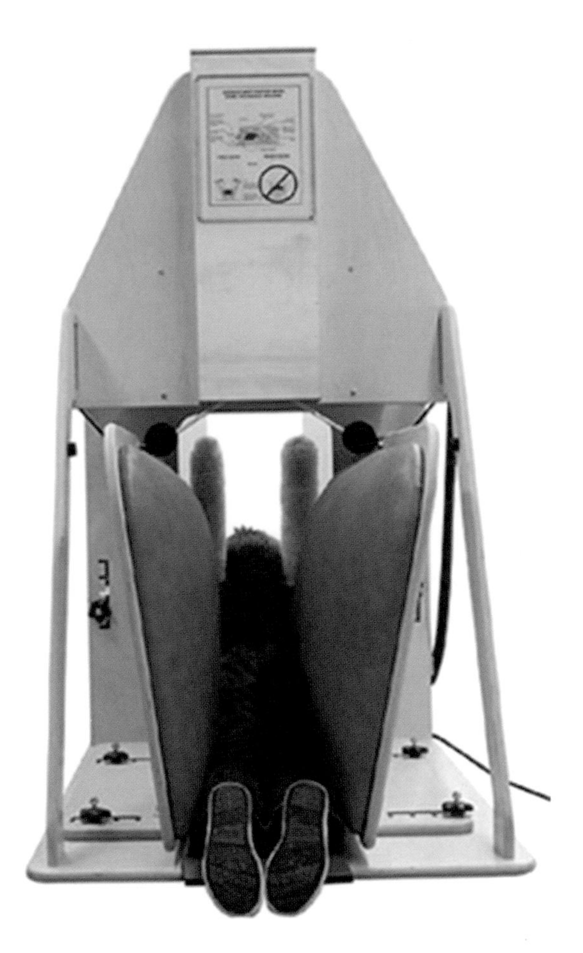

The Squeeze Machine is used for deep touch stimulation and produces a calming effect on hyperactive and autistic individuals.

In the history of inventors there are many remarkable women, but it is doubtful if there have been any as remarkable as the woman known simply as Temple Grandin. An inspiration to many families around the world, she has carved out two careers. One of the most accomplished writers and spokespersons about autism, she is also a hugely respected designer of handling facilities for livestock.

She was born Mary Temple Grandin in 1947 in Boston, Massachusetts, into an educated and wealthy family and earned a degree in psychology from Franklin Pierce College in 1970 and a master's in animal science from Arizona State University five years later. A doctorate in animal science from the University of Illinois at Urbana-Champaign followed.

A squeeze chute for holding cattle, horses, or other livestock, La Reata Ranch, Saskatchewan, Canada.

THE NERDY KID

Temple Grandin was never diagnosed with autism as a child and, although her mother reasoned that her atypical behavior was best explained by autism, it was not until she was in her forties that she received a diagnosis confirming that she was on the autism spectrum. Interestingly, when she was still a child, the general approach to people with autism was to institutionalize them, but, although Grandin's father supported that for his daughter, her mother fought against it.

Instead, she was given speech therapy from the age of 2 ½ years and her parents hired a nanny when she was 3 to play educational games with her for hours on end. She went to private schools that were sympathetic to the needs of a child like her but she has said that her time at junior high and high school were the worst times of her life as she was targeted as the "nerdy kid" by other students and subjected to bullying.

SELF-CONFIDENT AND OUTGOING

After expulsion from one school, she enrolled at Mountain Country School in New Hampshire, a school for children with behavioral problems. There she met William Carlock, a science teacher who had worked for NASA. He became a mentor to Temple Grandin, helping her to become more self-confident and outgoing. He encouraged her to develop an idea she had for the construction of what is now known as a "hug box" or "hug machine," although Grandin prefers to call it a "squeeze box."

She was 18 at the time and came up with the idea for the box after watching cattle on her aunt's Arizona ranch being confined in a squeeze chute for inoculation. Some of these animals, she noticed, calmed down after the pressure of the chute was applied. Disapproving of her machine, psychologists at her college tried to confiscate it but Carlock supported her and encouraged her to research how and why the machine seemed to work.

DEEP PRESSURE STIMULATION

The Hug Box is a deep-pressure device that is designed to bring calmness to hypersensitive people such as those with autism. It is made up of two side-boards measuring four by three feet that are hinged and have thick, soft padding. These form a V-shape and have, at one end, a control box and thick tubes that lead to an air compressor. The user lies or squats between the side-boards for as long as he or she wishes and, using pressure that is exerted by the air compressor, the side-boards move in and apply deep pressure stimulation over the body, soothing and calming the person inside.

Studies have proved that Temple Grandin's device does actually ease tension and anxiety in children with autism and it is used in some therapy programs. She admits that she continued to use it regularly to reduce anxiety until just a few years ago. "It broke two years ago," she told *TIME* magazine in 2010, "and I never got around to fixing it. I'm into hugging people now."

WE OWE ANIMALS RESPECT

But Temple Grandin's story does not end with the Hug Box. Following her master's in animal science, she worked as a consultant to large animal slaughterhouse companies, advising them how they could improve the quality of life for the animals before slaughter. Using her own personal experience with autism, she advocated that everything possible be done to remove items at the slaughterhouse that might cause the animals anxiety—fluttering flags, noise, shadows, and light being just a few of these things.

She talks of personally going through the chutes that animals are led through at slaughterhouses to better understand their experience. She has also invented a numerical scoring system for assessing animal welfare at slaughterhouses which has resulted in significant improvements in how animals are treated and eventually dispatched. She has said: "I think using animals for food is an ethical thing to do, but we've got to get it right. We've got to give those animals a decent life, and we've got to give them a painless death. We owe the animal respect."

Temple Grandin has written several books about autism and others about animal welfare. A film, starring the *Homeland* actor Claire Danes, has been made about her life. Today she teaches courses on livestock behavior at Colorado State University while continuing to advise the livestock industry on animal welfare.

Claire Danes and Temple Grandin attend the 2010 movie premiere of *Temple Grandin*.

GIULIANA TESORO

MORE PATENTS THAN ANY OTHER WOMAN
INVENTION: FLAME-RETARDANT FIBER

A researcher extinguishes a fire while wearing a fully protective flame-retardant suit.

Born into a Jewish family in Venice, Italy, Giuliana Cavaglieri (1921 – 2002) fell victim to the racial laws passed by Italy's fascist government which prohibited her from going to university. To further her education, she moved first to Switzerland, before traveling to the United States where she completed Yale University's graduate course in record time. At the age of just 21, she was awarded a doctorate in organic chemistry. She also married Victor Tesoro with whom she had two children.

During her long career as an organic chemist, Dr. Tesoro made many contributions to the fiber and textile industry, and was granted more patents than probably any other woman inventor. She has more than 125, most of which are associated with flame-resistant fibers. Her work in this area was vital as the use of organic polymers in fabrics became more common in the late 1970s. Finding a way to prevent these materials from burning was essential to their success and to consumer safety. She also discovered ways to stop the accumulation of static in synthetic fibers and created improved permanent press properties for textiles. She died in 2002 at the age of 81.

HALLDIS AALVIK THUNE

SAFE TRANSPORTATION OF THE INJURED

INVENTION: THE NECK-AID SUPPORT

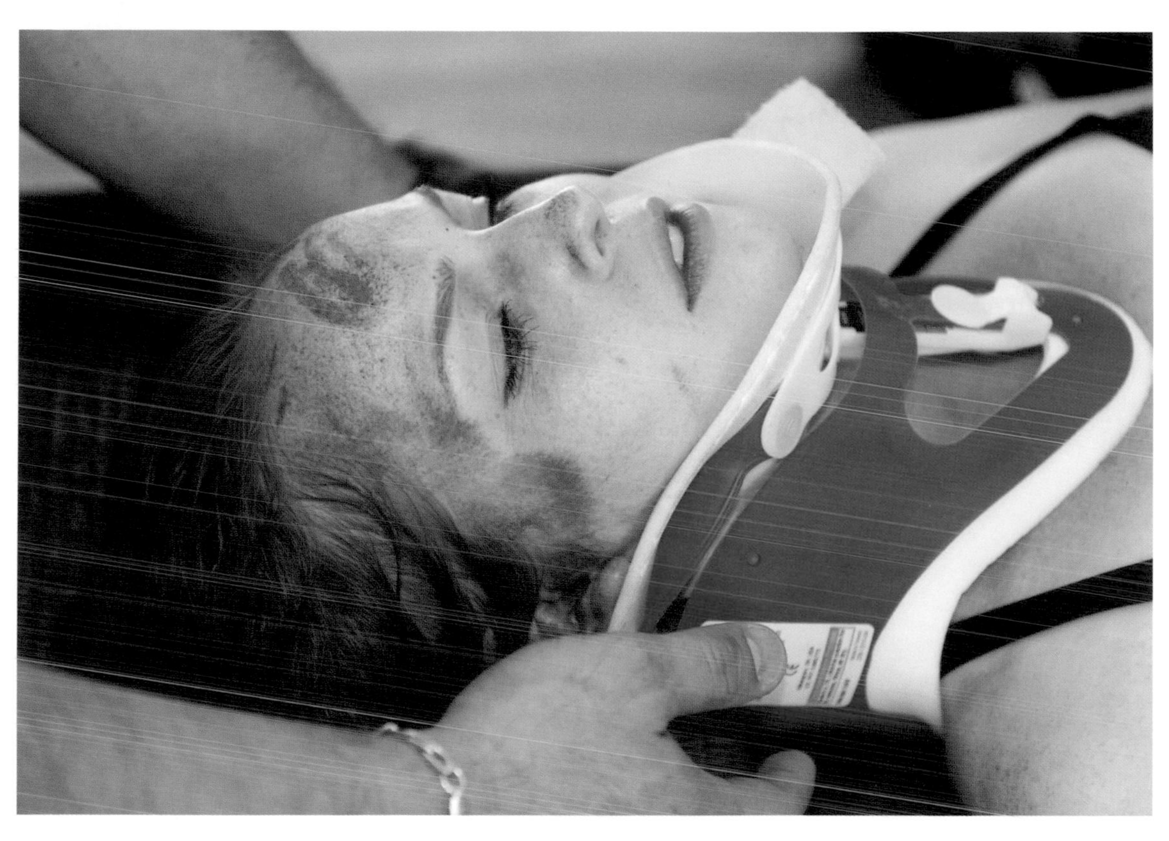

Paramedics fit a neck support to an injured patient.

Halldis Aalvik Thune was born in Norway in 1936, into a family of eight. The daughter of an inventor, she began to train as a nurse after first studying history and ethnology at university. In the mid-1980s, she invented the Neck-Aid and her life was never the same again.

Neck-Aid is a stiff support for the back and the neck that allows the safe transport and movement of accident and injury victims. It consists of wide cloth wrappings that stretch around and fasten in front of the torso of the injured person. They also fasten with Velcro across the forehead. Thus, the upper body and head are immobilized and the victim can be safely transported to the hospital for further evaluation and treatment.

The idea arose when she was a witness to a car crash in which one of the drivers was trapped behind the steering wheel. The grooved draining board next to her kitchen sink came into her head and she conceived of that kind of stiff, strong support which would be lengthened to the size of a body and placed under the injured person before they were moved. She left college and launched a company to market her invention which, made of strong plastic, hit the market in 1986, bringing her good sales and a number of awards.

EMMA LILIAN TODD

ON THE WINGS OF AN ALBATROSS
INVENTION: THE TODD BIPLANE

Emma Lilian Todd (1865 – 1937), a self-taught inventor with a love of flying, was probably the first woman ever to design an aircraft. It was a long way from where she started her working life, teaching herself to type to earn a living. Her first job, ironically, was at the United States Patent Office in Washington DC in 1888.

It was around 1903 that Lilian started working on flying machines. Inspired by airships she saw on a visit to London and Louisiana in 1904 and a sketch of an airplane in a newspaper in 1906, she exhibited her own design for a biplane at an air show at Madison Square Garden. It immediately made her a national sensation and secured the backing of a wealthy philanthropist, Olivia Sage, who provided her with funds (around $7,000) to build her airplane.

Lilian created her design having studied the wings of an albatross in the Museum of Natural History. "The wings or planes of my machine," she said, "are curved both lengthwise and crosswise, in order to deflect the air when it strikes the planes." The frame was of spruce and the muslin-covered wings were held together with piano wire. A two-seater powered by a modified Rinek motor, it was built in a large shed in Mineola, New York.

Finally in 1908, after several rebuilds, "The Todd Biplane" was ready to take to the air. A crowd gathered at the Garden City Aviation Field to see if the French pilot she had engaged, Didier Masson, could leave the ground. In fact, he not only left the ground, he flew for twenty feet, turned the plane and landed again safely.

Lilian was keen to pilot the craft herself so that she could fly it around the country, but when she applied for a pilot's license in 1909 her application was denied, probably because she was a woman. Newspaper reports confirmed her desire to fly but it never happened probably because the New York authorities would not relent. In fact, it is not known if she ever did fly her plane. Eventually, she took a job with Olivia Sage and gave the biplane to the State of New York. She died in 1937 at age 72.

Lilian Todd tests The Todd Biplane in 1908.

ANN TSUKAMOTO

UNDERSTANDING BLOOD IN CANCER RESEARCH

INVENTION: HUMAN STEM CELL ISOLATION

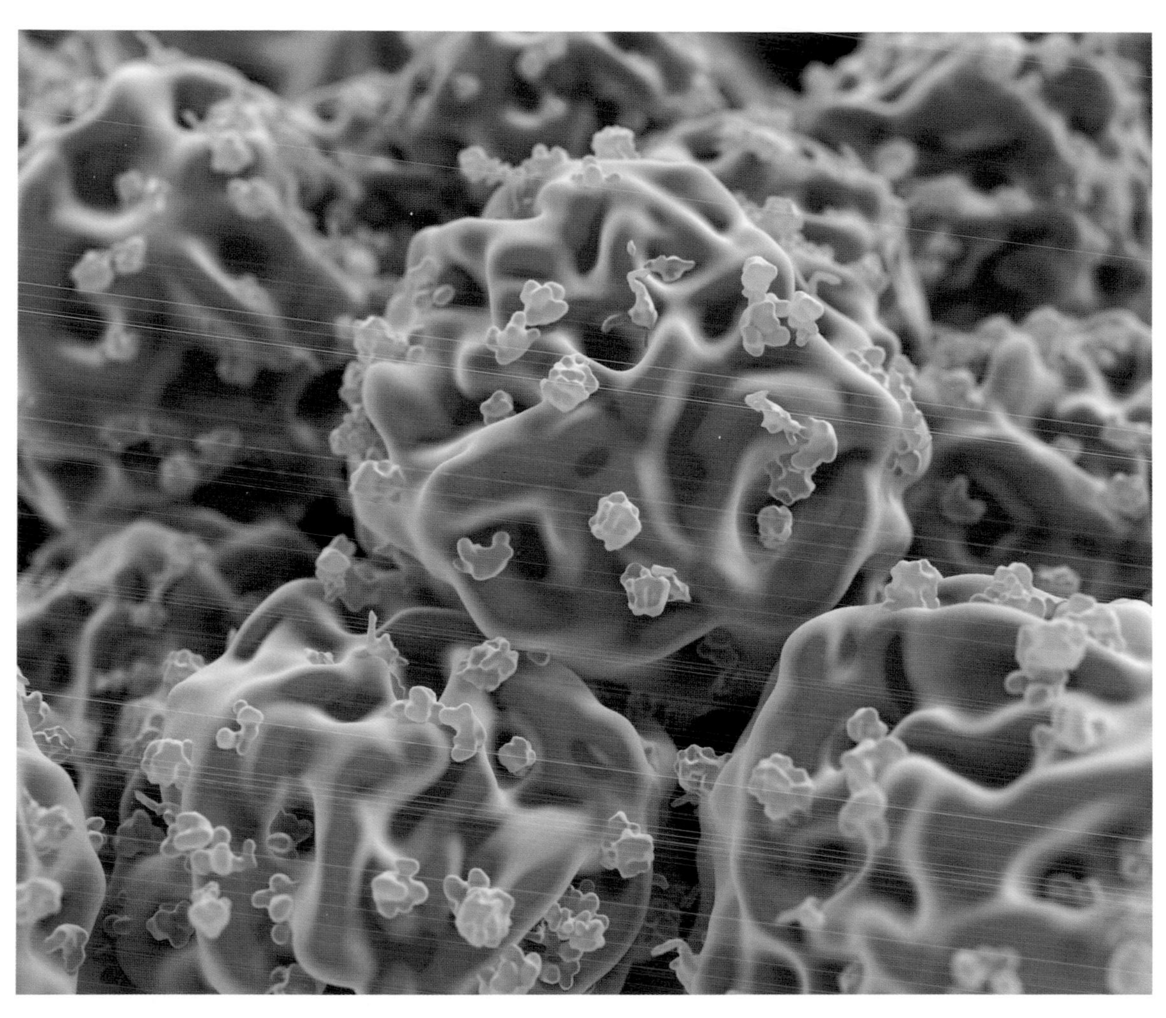

Stem cells seen through a scanning electron microscope.

Stem cells can be found in bone marrow and they are the foundation for the growth of red and white blood cells. In recent years, cancer research has focused a great deal of attention on them because how they grow and whether they can be produced artificially could prove to be highly significant in the eradication of cancer. In order to do this, stem cells had first to be isolated and the person who did this was American scientist Ann Tsukamoto (born 1952). She is an inventor on two issued US patents related to the human hematopoietic stem cell. She was awarded the co-patent for the process in 1990 and her work has led to a greater understanding of the blood systems of cancer patients and may hopefully lead one day to a cure for the disease. Ann Tsukamoto is currently directing further research in the areas of stem cell growth and cellular biology.

MADELINE M. TURNER

THE JUICE IS WORTH THE SQUEEZE

INVENTION: TURNER'S FRUIT PRESS

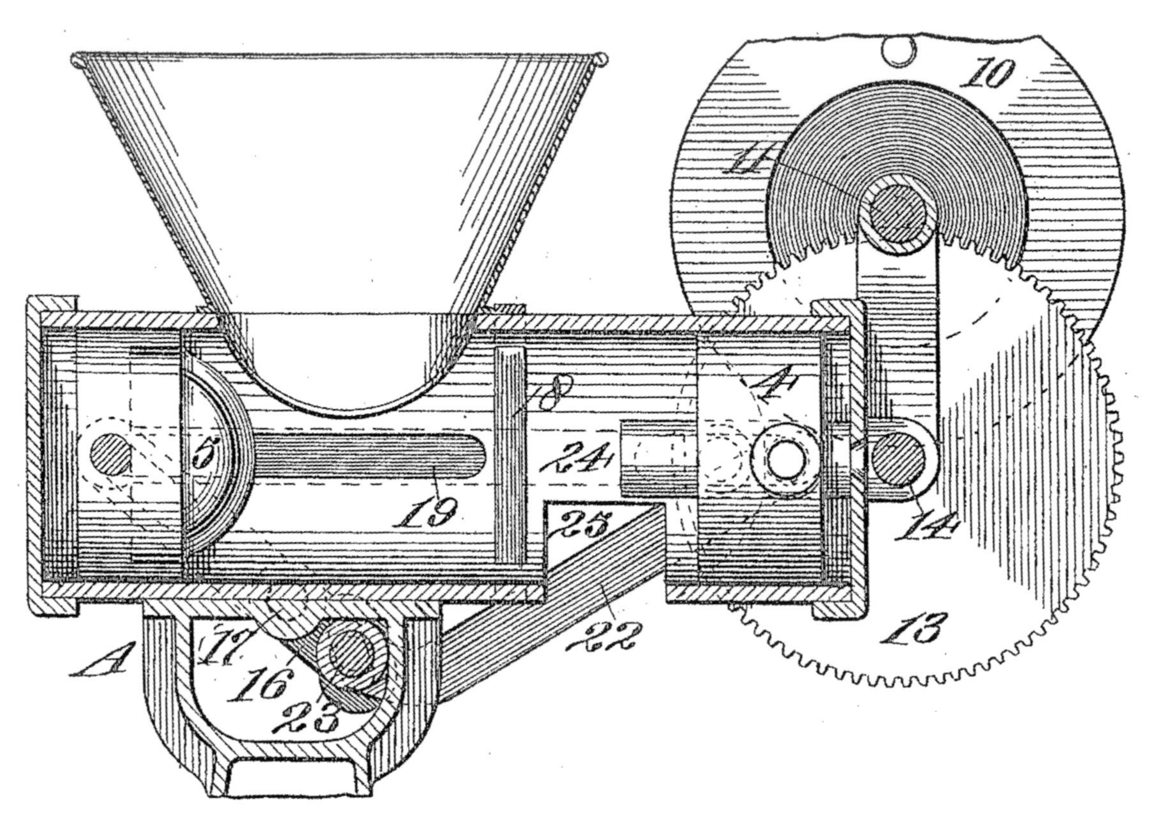

Madeline M. Turner's patent diagram for improvements in presses to extract the juice from fruits.

California-born African American inventor Madeline M. Turner was granted US Patent Number 1,180,959 on April 25, 1916, for a device that the Patent Examiner herself called "ingenious." It was a press for extracting the juice from fruit such as oranges or lemons.

The fruit was pushed into an opening and cut in half before being shifted between various plates until it had been completely juiced. The efficacy of her machines depended entirely on performing several tasks at the same time. Her invention was like an assembly line in a car factory, using a geared mechanism to move the fruit through the process. The fruit passed through stationery knives that cut it in half and then through the presser. The pulp dropped out through one opening and the juice through another.

Turner's Fruit Press was a far from simple device but it presaged the type of freshly squeezed juice machines that are now commonplace in the world fruit juice industry in the twenty-first century.

MARIE VAN BRITTAN BROWN

KEEPING AN EYE ON YOUR HOME
INVENTION: THE HOME SECURITY SYSTEM

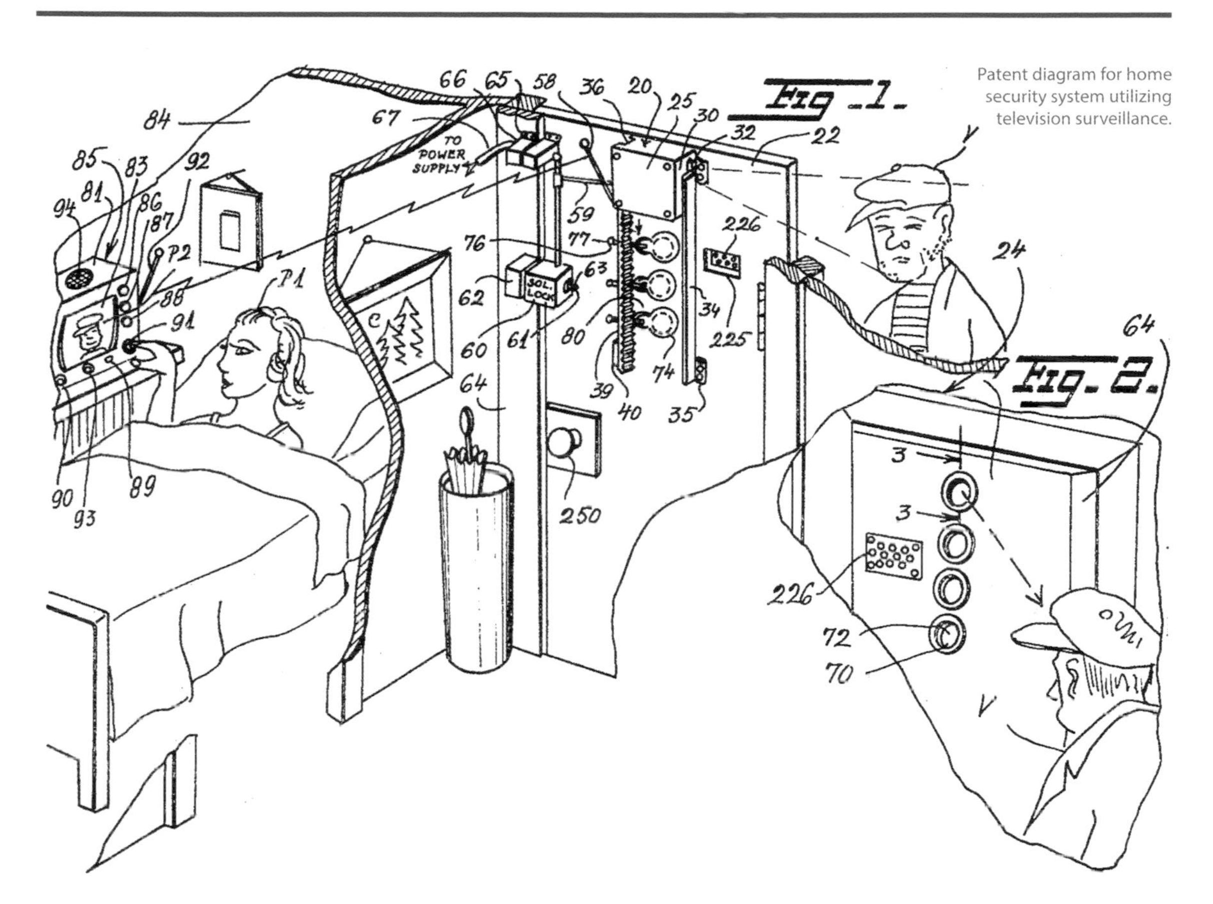

Patent diagram for home security system utilizing television surveillance.

New Yorker Marie Van Brittan Brown (1922 – 99) was an African American nurse who worked irregular hours as did her husband Albert, an electronics technician. Often she would be home alone and did not feel safe. She and Albert devised a clever home security system that was a forerunner of the CCTV home security technology in use today. The invention was inspired by the length of time police took to arrive after being called out by worried residents. She had the idea in 1966 and the patent was granted in 1969 for "a video and audio security system for a house under control of the occupant thereof. Occupant can see who is at the door ..."

It was a system for a motorized camera to display images on a monitor in the bedroom. There was a set of four peepholes and the camera could be remotely controlled to slide back and forth to each peephole to record what was happening outside. The images captured at the door were shown on the monitor using a radio-controlled wireless system. Furthermore, she could unlock any door remotely, the system could show the viewer inside who was at the door and, using a two-way microphone, they could be interviewed as they were standing there.

An additional feature was a button that could be pressed if the viewer was unhappy about the person presenting themselves at the door. This sent an alarm to a security firm, a member of a neighborhood watch group, or even a neighbor. If the person was to be allowed access to the house another button was pressed to unlock the door.

There is no record as to whether Marie's system ever was produced or whether she and her husband actually made any money from it, but they laid the foundations for a home security industry that will soon be worth $1.5 billion.

ADA HENRY VAN PELT

UTILIZING MOMENTUM
INVENTION: A DEVICE TO ELIMINATE FLYWHEELS FROM STEAM ENGINES

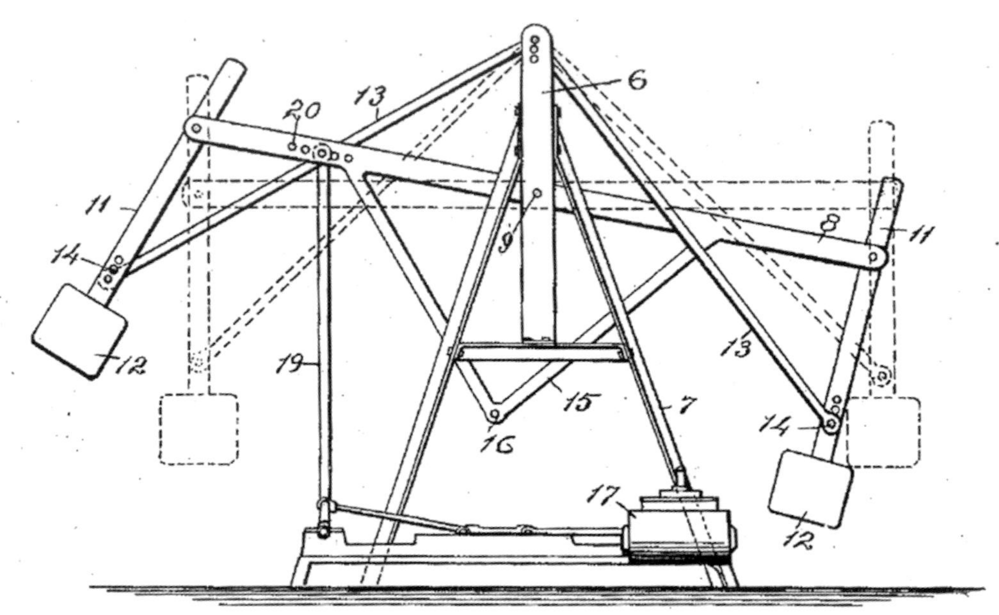

Patent diagram 1002610. An apparatus for utilizing momentum.

Playwright, temperance campaigner, magazine editor, and inventor Ada Van Pelt was quite a woman as demonstrated by the fact that she was still being granted patents at the age of 74.

She was born Ada Henry in 1838 in Princeton, Kentucky, the daughter of a banker. She married a soldier, Captain Charles E. Van Pelt, and after the Civil War the couple moved to Nebraska. Meanwhile, she wrote two plays that were produced in San Francisco, a Civil War drama, *The Quaker Sentinel* and a comedy, *The Cross Roads School.*

She also edited a temperance weekly entitled the *Pacific Ensign* but resigned in order to go on a lecture tour, presenting what she called "An Evening in California" during which she showed "stereopticon" views of the sights of that state and talked about her experience with the First Tennessee Regiment and the work of the Red Cross during the Spanish-American War.

Her foray into the world of inventing seems to have been met with incredulity. The reaction of the *Louisville Courier-Journal* of 1908 was typical:

> *... Mrs. Van Pelt's invention ... has been hoped for and sought in vain by mechanical engineers all over the world for more than fifty years ...*

Mrs. Van Pelt herself described her Apparatus for Utilizing Momentum in US Patent No. 1002610, dated September 5, 1911:

> *It is my purpose to dispense with the ordinary fly wheel and substitute therefore a weighted oscillating beam, which is provided with swinging pendants, so that I am able to obtain the benefit of the weighted bar as well as the pendulous motion which is set up by the oscillations or vibrations of the pendulum when properly connected up with an engine or other source of power.*

Her first invention had been a combination lock that was adapted for use by the United States Government on letter boxes, and she later devised an improved style of letter box that was adopted by the Post Office Department and used in cities in the east of the United States. Her last invention was a water purifier. Ten years later in 1923, she died at age 84.

LADY ANN VAVASOUR

UNEARTHING SIX ACRES A DAY
INVENTION: MACHINERY FOR TILLING LAND

If it had ever been made, Lady Vavasour's "enormous machine" may have resembled this modern-day Tree Muncher Land Clearing Machine.

Lady Ann Vavasour was granted a patent in 1842 for what was described as "Machinery for Tilling Land." A member of the Anglo-Irish gentry, her machinery came about after she had observed workers in an Italian vineyard break up the soil around the base of the vines with long, bent forks. She conceived of an implement that could work in the same way, thus taking some of the hard work out of plowing.

She crafted her first model in the German town of Rippoldsau and then involved a good engineer, her objective being to devise a machine that "could dig and pulverize," as she put it, "six acres at least of land a day." It would be to a depth of ten inches and would use two or at most three horses. She was firmly of the opinion that it would be more efficient than any other plow. "My implement will leave [the soil] open and broken up, so that the rain can sink in."

Unfortunately, at the 1842 Bristol Agricultural Show, Lady Vavasour's invention proved something of a disappointment. Described in the *Journal of the Royal Agricultural Society* as an "enormous machine," the earth stuck to the teeth of the barrel, closing up the spaces in between, and the plow acted more as a roller of the ground than a plow.

We cannot be certain what caused this malfunction. It could have been that the ground was just too wet or the design, as interpreted by the manufacturer of this prototype, may have been faulty. Nonetheless, to have persevered and got it that far was testament to the determination and enthusiasm of Lady Vavasour and was no mean feat for a woman at that time.

JEANNE VILLEPREUX-POWER

EXPERIMENTING WITH AQUATIC ORGANISMS
INVENTION: THE AQUARIUM

The shark tank at L'Oceanogràfic, the largest aquarium in Europe, Valencia, Spain.

The eldest child of a shoemaker, Jeanne Villepreux was born in France in 1794. Her upbringing was in the rural countryside, and at the age of 18, she set out on foot for the 400 kilometer journey to Paris. On her arrival, she found work as a society dressmaker and in 1816, she became famous for creating Princess Marie-Caroline's royal wedding dress. Two years later, Jeanne married an English merchant named James Power and moved with him to Sicily.

It was in Sicily that Jeanne Villepreux-Power discovered the passion of her life, marine biology. She began to learn about the sea as she explored the coast of Sicily, recording its flora and fauna and collecting fossils, minerals, and shells. Soon, she began to focus on the creatures of the sea, making astonishing discoveries. She was the first person to work out how the pelagic octopus *Argonauta Argo* creates the thin shell casing that coils around it and how it reproduces.

Probably the most significant outcome of this research, however, was her invention of the aquarium. She needed to study these aquatic creatures close-up and the only way was some kind of watery enclosure. She devised three different types of aquaria. The first was a simple glass one as persists to this day. A second type was also made of glass but was surrounded by a cage so that she could submerge it in the sea to study molluscs. The third was a larger cage that was for studying larger molluscs further out to sea. She published her results in two books, one in 1839, the other three years later.

Jeanne and her husband left Sicily in 1843, living from then on in London and Paris. Tragically, the ship that was carrying all her collections, records, and drawings sank on the way to London and all was lost. Nonetheless, she continued to write and publish but undertook no further research. She fled the siege of Paris by the Prussians in the winter of 1870 and returned to her childhood home of Juillac where she died not long afterward.

RUTH GRAVES WAKEFIELD

BAKING A NATIONAL PHENOMENON

INVENTION: THE CHOCOLATE CHIP COOKIE

Bake up extraordinary chocolate chip cookies in no time. They're a cookie-jar favorite!

It is often said that chocolate chip cookies were no more than one of those happy inventing accidents. Rumor has it that Ruth Wakefield had actually expected the chunks of Nestlé's semi-sweet chocolate bar to melt as she stirred them into her chocolate cookie dough. The reality was that they did not melt but that outcome was entirely intentional. "We had been serving a thin butterscotch nut cookie with ice cream," Ruth said. "Everybody seemed to love it, but I was trying to give them something different. So I came up with the Toll House cookie."

The Toll House Inn was a restaurant in Whitman, Massachusetts, run by Ruth and her husband. She was a good cook and the place was soon known for its tasty lobster dinners and wonderful desserts. She published a cookbook, *Toll House Tried and True* with recipes which proved immensely popular in the 1930s. In 1938, the Toll House Cookie recipe was included.

During World War II, US soldiers from Massachusetts received care packages from home with Toll House Cookies inside that their wives had baked using Ruth's recipe. They shared them with soldiers from other parts of the country. Before too long, soldiers from all over America were sending letters home asking for Toll House Cookies to be sent and Ruth was inundated with letters asking for the recipe. The chocolate chip cookie had become a national phenomenon.

At the same time as the chocolate chip cookie was taking off, so too were sales of the Nestlé chocolate bars and Ruth came to an agreement with Nestlé that they could use her recipe and the Toll House name in return for just one dollar and a lifetime supply of Nestlé chocolate bars. They started production of the chocolate chips for the cookies and printed her recipe on the packaging. The chocolate chip cookie became the most popular variety of cookie in America and remains so to this day.

Toll House Cookies tin from the 1930s.

MADAM C.J. WALKER

THE GREATEST BENEFACTRESS OF HER RACE
INVENTION: BEAUTY AND HAIR PRODUCTS

Madam C.J. Walker's company delved into products like this "skin brightener" to stay afloat in the 1930s.

Sarah Breedlove, a.k.a. Madam C.J. Walker, was probably the first self-made female millionaire in America. During her lifetime she was certainly one of the wealthiest African American women in the country, and one of the most successful African American business owners ever.

Born in 1867 in Delta, Louisiana, Breedlove's parents and older siblings were slaves but the Emancipation Proclamation meant that she was the first in her family to be born free. By the age of 7, however, she was an orphan and at 10, she was in domestic service in Vicksburg, Mississippi. In 1882, at the age of 14, she married Moses McWilliams who died five years later, leading her to re-marry. It did not go well and in 1905 she left her husband and moved to Denver, Colorado.

BECOMING MADAM C.J. WALKER

Her third marriage in 1906 was to a newspaper advertising salesman, Charles Joseph Walker and it was at this time she began to be known as Madam C.J. Walker. When Charles died in 1912, Sarah moved with her daughter Lelia, daughter of Moses McWilliams, to St. Louis, Missouri, and worked in a laundry, earning very little but determined to provide an education for Lelia.

Like many black women of her time, Sarah suffered from severe dandruff and scalp problems. Some women had issues with baldness due to problems with their skin and the use of chemicals that were in products for clothes-washing and for washing the hair. Naturally, poor diet, not washing enough and a lack of basic amenities such as heating, electricity, and proper plumbing also contributed.

One of the reasons Sarah had relocated to St. Louis was to be near her four brothers who lived there. They were all barbers and from them she learned about hair care. She also began working for Annie Pope-Turnbo, an African American hair-care entrepreneur who owned the Poro Company. Sarah worked selling her product, but later became a deadly business rival.

SPREADING THE NET WIDER

Moving back to Denver, Colorado, she continued selling Poro products but the success she was enjoying with these products suggested to her that she could probably do better with her own. She created products specifically for African American women and spent a year on the road in the South, building up a client-base for a mail-order business.

She opened a shop in Pittsburgh, Pennsylvania, and by 1908 she had trained hundreds of sales agents. Establishing her headquarters in Indianapolis, she began to create a reputation for philanthropy, donating $1,000 to the Young Men's Christian Association who were trying to build a facility in a black neighborhood.

The Madam C.J. Walker Manufacturing Company of Indiana was founded in 1910, and she provided all the funding for the business. As she trained increasing numbers of agents, she also traveled to the Caribbean and Central America, spreading the net wider for sales of her products. In those regions she also trained new agents.

FUNDING ANTI-LYNCHING PROGRAMS

1916 saw the establishment of the Madam C.J. Walker Benevolent Association. Good publicity, she figured, was good for business. But she really was very committed to improving the lives of African Americans, funding anti-lynching programs organized by the National Association for the Advancement of Colored People and the National Association of Colored Women.

She moved to New York in 1916 to be close to her daughter who was living there, purchasing land at Irvington-on-Hudson and building a $100,000 mansion. But a year later she was diagnosed with high blood pressure and kidney problems and told to slow down. Characteristically, she ignored her doctors and took on more.

Eventually, however, she was unable to travel. It did not stop her joining a delegation of Harlem leaders who went to Washington DC to argue for equal rights for black soldiers fighting in World War I. Madam C.J. Walker died in 1919, leaving the majority of her estate to charity.

THE EMANCIPATION PROCLAMATION

The Emancipation Proclamation was an Executive Order that was issued by President Abraham Lincoln on January 1, 1863, during the American Civil War. By this order, the status of more than three million people in ten designated states of the American South was changed from enslavement to freedom. By running away or if the area they lived in was captured by Union soldiers, the slaves there were freed. There was no compensation for slave owners and neither did it outlaw slavery. It did not even grant citizenship to freed slaves. But with it, the eradication of slavery became an explicit war goal.

Original tins of Madam C.J. Walker's Glossine, Scalp Ointment and Wonderful Hair Grower from 1925.

INVENTING VOTES FOR WOMEN

Women have been fighting to win the right to vote since the middle of the nineteenth century. Curiously enough it was the rugged western states of America that led the way, with Wyoming and Utah granting women the right to vote in 1869 and 1870. It was said the inventive authorities there hoped to attract more single marriageable women to the regions. Similarly short of female members of the population, New Zealand and Australia granted votes for women in 1893 and 1902 respectively. The first European country to introduce women's voting was Finland in 1907 followed by Norway in 1913. Most independent countries granted votes for women in the interwar years. The end of World War I heralded a time of great change, and the belief that women were somehow temperamentally unfit to vote was seen as a notion from another time. Canada granted votes for women in 1917, Britain and Poland in 1918, and the United States in 1920.

Emmeline Pankhurst (1858 – 1928) was the influential leader of the British suffragette movement who fought an increasingly bitter and violent protest campaign to help women win the right to vote in Britain. In 1999, *Time* magazine commented that Emmeline Pankhurst "shaped an idea of women for our time … from which there could be no going back," and listed her as one of the 100 Most Important People of the Twentieth Century.

(Facing page) Movie poster for *Suffragette* (2015), with Meryl Streep starring as Emmeline Pankhurst.

CAREY MULLIGAN HELENA BONHAM CARTER BRENDA GLEESON ANNE-MARIE DUFF AND MERYL STREEP

London, 1912.
Before women had a voice, they had to fight for their freedom.

SUFFRAGETTE

A FILM BY SARAH GAVRON

IN THEATRES 31 DEC

MARY WALTON

A WOMAN'S BRAIN DID THE WORK
INVENTION: POLLUTION-REDUCING DEVICES

The Industrial Revolution brought many major benefits but its main disadvantage was the pollution it created, factories belching smoke from their chimneys and the new railways making noise on an unprecedented level. Mary Walton introduced methods of dealing with both smoke and noise at a time when environmental issues were far down the list of concerns.

She had been working on smoke reduction around 1879 when she came up with a method of reducing the dangers caused by the smoke that poured from factory chimneys. That year she was granted a patent for a system whereby the emissions were deflected from the top of the chimneys and drawn down into water tanks where they were stored until they could be disposed of by being flushed into the sewage system.

Two years later another patent came her way. This time she was dealing with the other blight of nineteenth-century life—noise. Elevated railways were being built in the bigger American cities, producing a huge amount of noise all day as the metal train wheels rattled over the metal rails. Living in Manhattan, Walton had first-hand experience of such noise. In the basement of her house she built a model railway and began work on noise-reduction ideas.

Her solution to the constant rattling and clanging of train wheels on track was to frame the rails in a wooden box-type construction. This was painted with tar, lined with cotton wool and filled with sand. The vibrations and sounds were deadened by the sand and cotton wool. After successful trialing, New York City's Metropolitan Railroad bought the rights from Walton for $10,000. She was hailed as a hero, her success noted by the *Woman's Journal* twenty years later:

> *The most noted machinists and inventors of the century had given their attention to the subject without being able to provide a solution when, lo, a woman's brain did the work.*

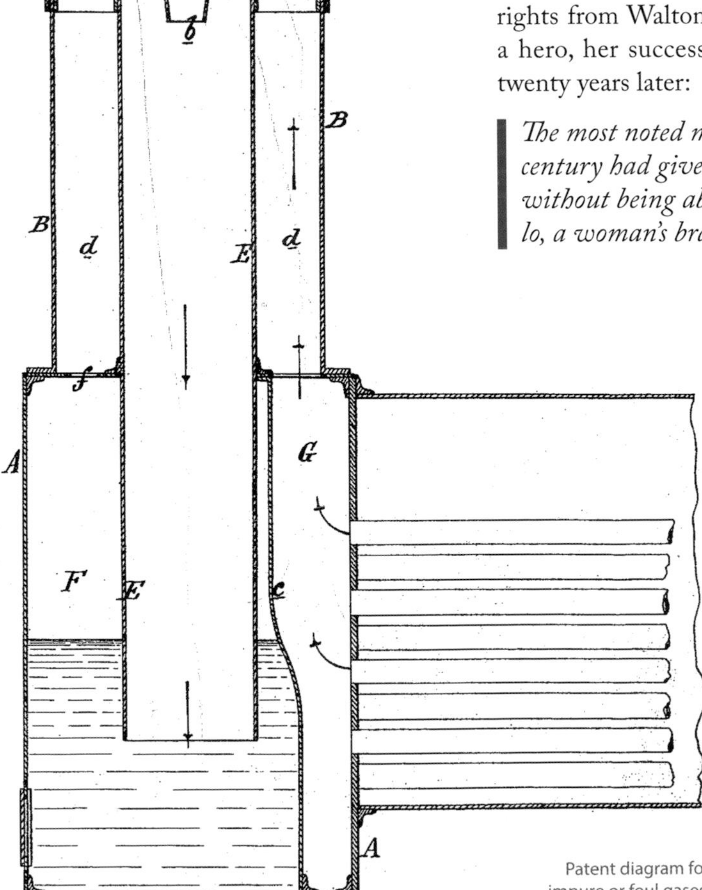

Patent diagram for preventing the escape of sparks, ashes, and impure or foul gases or vapors from the chimneys of locomotive-engines and from other chimneys into the atmosphere.

JUSTINE JOHNSTONE WANGER

THE GIRL WHO OWNED BROADWAY

INVENTION: INTRAVENOUS DRIP THERAPY

Intravenous drip therapy in the hospital.

Broadway, the Ziegfeld Follies, and the Folies Bergères would at first sight be unlikely places for the training of a woman who helped to invent one of the most significant techniques in medical history, but that is exactly how Justine Wanger began her working life.

She was born Justine Johnstone in Hoboken, New Jersey, in 1895 and by the age of 15 had made her *entrée* into show business, appearing as a mannequin for Madame Sherry in New York City. A year later, she was performing with the Folies Bergères in Paris followed by two years of vaudeville, before she took on a role in a Broadway musical, *Watch Your Step*. She was in the Ziegfeld Follies of 1915 and appeared in two other Broadway musicals before getting a lead role in *Over the Top* in 1917. She became a big star and was known in the press as "the girl who owns Broadway."

But being intelligent as well as talented, Justine wanted a little more meaning in her life. After more theater and starring in a few silent films, she married the American film producer Walter Wanger. Her marriage gave her the opportunity to study science, and she enrolled at Columbia University. She became a research assistant to Dr. Samuel Hirshfield and his colleague Dr. Harold T. Hyman, who were studying speed of injection on patients.

It was thought that injections being carried out too rapidly was probably the cause of some deaths, a phenomenon for which the two doctors coined the term "speed shock." Justine Wanger's job was to undertake most of the research. Eventually, they confirmed that, indeed, the injection speed was critical to the success of intravenous treatments and transfusions.

Even if harmless material was injected too fast, the patient could die. However, toxic drugs could be delivered intravenously if done slowly over a period of some hours. Wanger, Hirshfield, and Hyman had invented the modern intravenous drip method of drug delivery that is now used in hospitals and medical facilities around the world. When the report was published, Justine Wanger, the girl who owned Broadway, was credited as one of the inventors.

ADELINE DUTTON TRAIN WHITNEY

LEARNING YOUR ABC
INVENTION: WOODEN ALPHABET BLOCKS

The child had to select the correct wooden blocks from the set in order to construct the individual letters of the alphabet.

Alphabet blocks (cubes with a letter printed on each side) have long been a useful tool for parents teaching their children the alphabet and how to read. They also help to stimulate the imagination and in their simplest use, as a building toy, are good for developing coordination. This simple but multipurpose toy has existed in some format since probably around 1693. The English philosopher John Locke wrote that "dice and play-things, with the letters ... teach children the alphabet by playing." They are called "building bricks" in the 1798 book *Practical Education* by Maria and R.L. Edgeworth while in 1820 in the United States they were described as "multi-colored blocks" and as "terracotta toy blocks" in England in 1850.

The alphabet blocks we know and love today were patented in 1882 by Adeline Dutton Train Whitney, the well-educated scion of a wealthy family who married at 19 and had four children. Writer of more than twenty books, mainly for young girls, Whitney was a champion of conservative values, vehemently opposed to women's suffrage and convinced that a woman's place was very much in the home.

MARGARET A. WILCOX

DRIVING IN COMFORT
INVENTION: THE CAR HEATER

Modern-day car heater controls.

Female mechanical engineers were rare in the nineteenth century but when they did come along they were usually pretty good, and Margaret A. Wilcox was no exception to that rule. Born in Chicago in 1838, she always wanted to invent something meaningful. When you climb into your car on a frosty morning and reach for the button that switches on the car heater so you can drive in comfort, you should be grateful that she succeeded. Her idea, like all the great ones, was simple. Warm air from the vehicle's engine was deflected into the car interior, warming the occupants. She received the patent for this innovation in 1893 and it remains the basis for car heaters to this day.

Wilcox invented a number of other things, including a combined cooking and hot-water-heating stove, and a combined clothes and dishwasher stove. This was based on a semi-circular tank inside which a cradle was rocked by a crank handle. Rubber mats served as a washboard for the clothes and these were taken out when dishes were being washed. The hot water was supplied by a gas boiler.

FLOSSIE WONG-STAAL

VIROLOGIST AND LIVING GENIUS
INVENTION: CLONING HIV

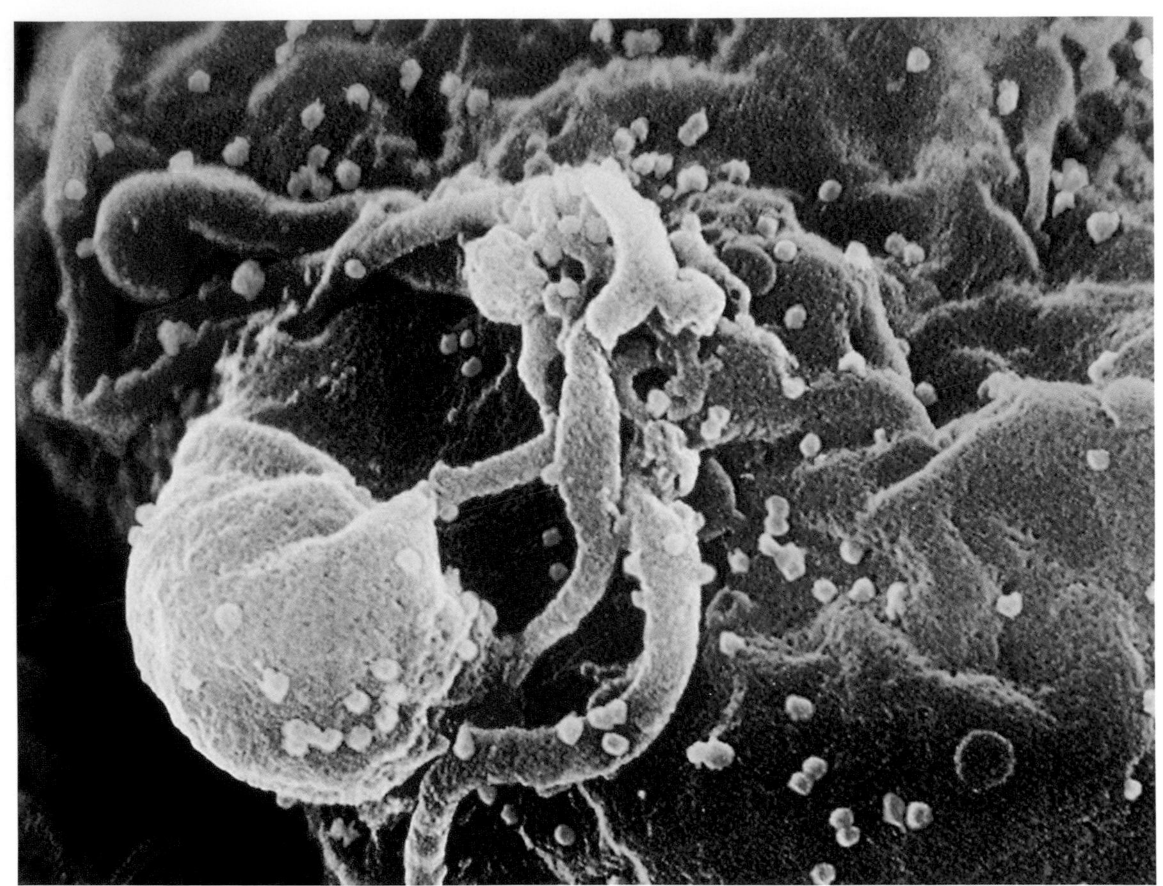

Scanning electron micrograph of HIV-1 budding (in green) from cultured lymphocyte.

Flossie Wong-Staal is a Chinese American virologist and molecular biologist. Born in 1947 in Guangzhou, China, as Yee-Ching Wong, she fled with her family to Hong Kong when the Communists came to power. At school there, she performed particularly well in science and was encouraged to go to the United States to further her education.

At the age of 18, therefore, she enrolled at the University of California at Los Angeles, studying for a Bachelor of Science degree, specializing in bacteriology. She followed this with a PhD in molecular biology in 1972. At the University of California, San Diego, she undertook post-doctoral research before leaving to work with the leading American biomedical researcher, Robert Gallo, at the National Cancer Institute.

She was part of the team that identified HIV as the cause of AIDS and two years later she was the first scientist to clone HIV and determine the function of its genes, a major step in proving that HIV is the cause of AIDS. This mapping also made it possible to develop blood tests for HIV. Wong-Staal was named top woman scientist of the 1980s by the Institute for Scientific Information and in 2007, *The Daily Telegraph* named her as number 32 of the "Top 100 Living Geniuses."

NANCY FARLEY WOOD

ATOMS FOR PEACE
INVENTION: THE IONIZING RADIATION DETECTOR

Firefighters wearing anti-radioactive suits and using ionizing detectors at an accident crash site where radioactive materials have been involved.

From a farming upbringing to the Manhattan Project is quite a journey but it is the one made by the remarkable Nancy Farley "Nan" Wood. The farm was located at La Monte, Pettis County, Missouri, where she was born in 1903, the second of four children. Her education began in a one-room schoolhouse but she was very bright and her family moved to central Missouri so that she could pursue her education at college. Graduating from the Warrensberg Teacher's College, she taught high school mathematics and physics before gaining an MA in education from the University of Chicago in 1927.

When America entered World War II in 1941, she was teaching calculus to US Navy personnel in Chicago while her husband stayed home to look after their children, a set-up that was very unusual at the time. Toward the end of the war, she was recruited by the FBI to work on the Manhattan Project, the US government's research project that was building a nuclear bomb. It was while working there that she and John Alexander Simpson designed and developed radiation detectors that detected ionizing particles, such as those emitted by nuclear decay, cosmic radiation or reactions in a particle accelerator. This research was carried out in the instrument division of the University of Chicago Metallurgical Laboratory.

In 1949, Nancy Wood founded the N. Wood Counter Laboratory in Chicago, a company with the aim of developing and manufacturing gas-filled gamma radiation detectors and neutron radiation detectors for use in laboratories and universities where research was being carried out into the peaceful use of atomic energy. It remained in business for over fifty years before Wood sold it to her daughter. Her detectors have been used in satellites by NASA.

Nancy Wood, who died in 2003 at age 99, was an independent, strong-minded woman, and one of the first feminists. In 1969, she even picketed United Airlines because women were not allowed to fly on executive flights between New York and Chicago.

CHIEN-SHIUNG WU

THE FIRST LADY OF PHYSICS
INVENTION: THE WU EXPERIMENT

Chien-Shiung Wu working in her laboratory at New York's Columbia University, USA, where she became physics professor in 1957.

Chien-Shiung Wu, "The First Lady of Physics," was born in Jiangsu province, China, in 1912 and was educated at one of the first schools in China to admit girls. In 1930, she enrolled at Nanjing University where, inspired by the life of Marie Curie, she studied physics.

Graduating with honors at the top of her class in 1934, she taught at the university for a year, before deciding to pursue her studies and research in America. In 1936 at the University of California, Berkeley, she met the pioneering American nuclear scientist, Professor Ernest Lawrence who had built the first cyclotron. He and another Chinese student, Luke Chia-Liu Yuan, encouraged her to stay and work at Berkeley.

She completed her PhD in 1940 and in 1942 married Luke Chia-Liu Yuan. She also accepted a post at Princeton University, the first woman to join the faculty. In 1944 she joined the Manhattan Project at Columbia University where she invented a process of enriching uranium ore to be used as fuel for atomic bombs. After leaving the project in 1945, Wu became the leading expert in beta decay and weak interaction physics at Columbia University.

In the mid-1950s, the research of two male theoretical physicists, Tsung-Dao Lee and Chen-Ning Yang, into the "law of conservation of parity" persuaded them that parity was conserved for electromagnetic interactions and for strong interaction. But the theory had never been properly tested for weak interactions. To test the theory, they turned to Wu for help in setting up suitable laboratory conditions because of her great expertise in weak interaction physics.

The experiment disproved the hypothetical "law of conservation of parity" for weak nuclear interactions and become known as the Wu Experiment. But controversially, it was the two male scientists who were awarded the 1957 Nobel Prize in Physics for this ground-breaking work, and Wu was excluded. It was a fate suffered by many female scientists at the time.

Chien-Shiung Wu was named Scientist of the Year in 1974 by *Industrial Research Magazine* and in 1976 was the first woman to serve as president of the prestigious American Physical Society. The First Lady of Physics died at age 84 in 1997 and was posthumously admitted into the American National Women's Hall of Fame.

ROSALYN SUSSMAN YALOW

THE MOST VALUABLE ADVANCE EVER INVENTED

INVENTION: RADIOIMMUNOASSAY

Dr. Rosalyn Yalow at her Bronx Veterans Administration Hospital, October 13, 1977.

The Nobel Prize committee, when making the 1977 award in Physiology and Medicine, described radioimmunoassay (RIA) as "the most valuable advance in basic research directly applicable to clinical medicine made in the past two decades." In RIA, radioisotopes and the body's immune response (creating antibodies that are specific to a given antigen) are used for measurement. It is one of the most effective techniques that has ever been developed for the measurement of substances such as hormones in human blood and tissues. For instance, doctors could, using RIA, for the first time measure the amount of insulin in the blood of a diabetic. It has since been applied to hundreds of substances all of which had been too minute to be detected beforehand.

RIA was the brainchild of Dr. Rosalyn Yalow (1921 – 2011) and Dr. Solomon A. Berson (1918 – 72) and it won Dr. Yalow a Nobel Prize in Physiology or Medicine five years after Berson died, the Nobel Prize being confined to the living. She was only the second woman to win the prize in that particular category. Both Yalow and Berson had refused to patent their invention despite its huge commercial value. Significantly, Rosalyn Yalow had a sign hanging in her office at the Veterans Administration Hospital in the Bronx that read: *"Whatever women do, they must do twice as well as men to be thought half as good. Luckily this is not hard."*

RACHEL ZIMMERMAN BRACHMAN

SPEAKING WITHOUT SPEECH
INVENTION: THE BLISSYMBOL PRINTER

Space scientist and inventor, Rachel Zimmerman Brachman was born in London, Ontario, in 1972. Her mother was a computer programmer and always encouraged her to learn more about science. At age 11, she became interested in Louis Braille and Helen Keller and their non-traditional methods of communication. In the school library she found a book entitled *Blissymbolics: Speaking Without Speech* which explains how cerebral palsy sufferers communicate using the Blissymbols method of pointing to images on a board. It was a laborious process, since the poor muscle control that comes with cerebral palsy made it difficult to point to the right image.

In 1984, at the age of 12, Rachel developed a software program to put the symbols onto a special computer touch pad. When the user chose a symbol, the "Blissymbol Printer," as she called her invention, converted the image to written English on a small screen, making it simpler for users with physical disabilities to communicate. It was a ground-breaking invention, a long time before people had email or internet access, and twenty years before iPads were invented.

Now a space scientist, Rachel works at NASA Jet Propulsion Laboratory where she adapts NASA ideas for those with disabilities and is a Solar System Ambassador.

Rachel Zimmerman Brachman and her Blissymbol printer.

WHAT ARE BLISSYMBOLS?

Blissymbols were invented by Charles K. Bliss, a citizen of the Austro-Hungarian Empire, while he was living in Shanghai and in Sydney between 1942 and 1949. His aim was to create an easy-to-learn language that would enable different linguistic communities to communicate with each other. He established an ideographic writing system called Semantography that consisted of several hundred basic symbols that each represented a concept. When these symbols were used together they could generate new symbols that represented new concepts. Unlike most of the world's writing systems, blissymbols do not correspond to any spoken language.

It is easier to learn to read and write blissymbols than it is to learn to read and write your own native language. As can be seen from the samples below, memorability and simplicity is what makes the language easy to learn. Even though each blissymbol can be learned more quickly than a word of a foreign language, the huge variety of unique symbols can sometimes appear overwhelming.

WHAT'S THE DIFFERENCE BETWEEN EMOJIS AND EMOTICONS?

These days the internet has greatly changed the way people communicate. Since body language and verbal intonation do not translate in our online messages, two new-age hieroglyphic languages have come to prominence to help convey our moods: emoticons and emojis.

Not to be confused with each other, emoticons are standard punctuation marks, letters, and numbers used to create pictorial icons such as the smiley face :-) and the frowning face :-(. Whereas the more recent emojis are cartoon faces with various expressions, as well as objects, animals, food, mathematical symbols, and many many more …

It was Apple intern Angela Guzman and her design mentor Raymond who were jointly responsible for creating Apple's first set of emoji designs in 2008 for which they were awarded a US patent. It is hard to believe that people did not even know what these tiny illustrations were less than 10 years ago.

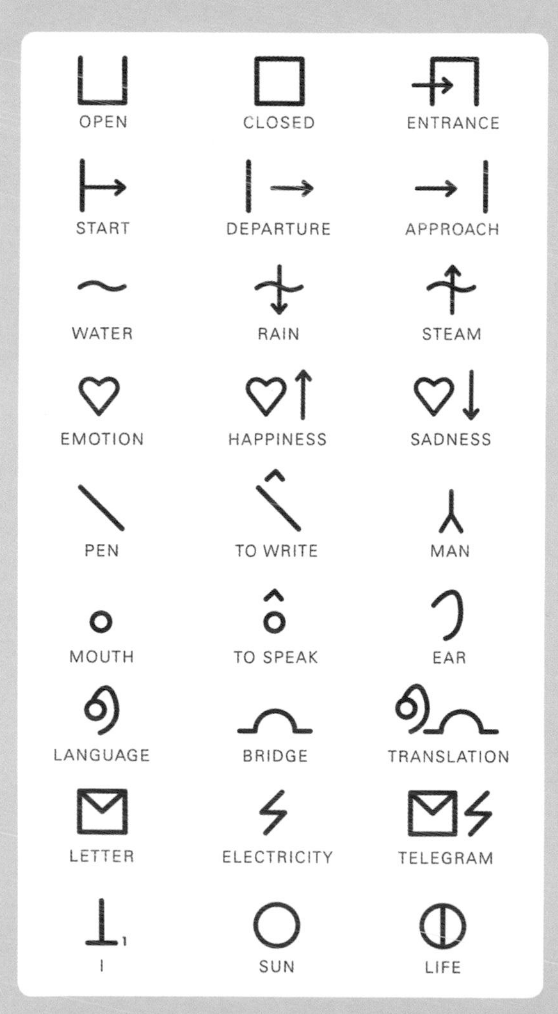

A sample of the huge variety of blissymbols.

A sample of the now-familiar emoji icons.

INDEX

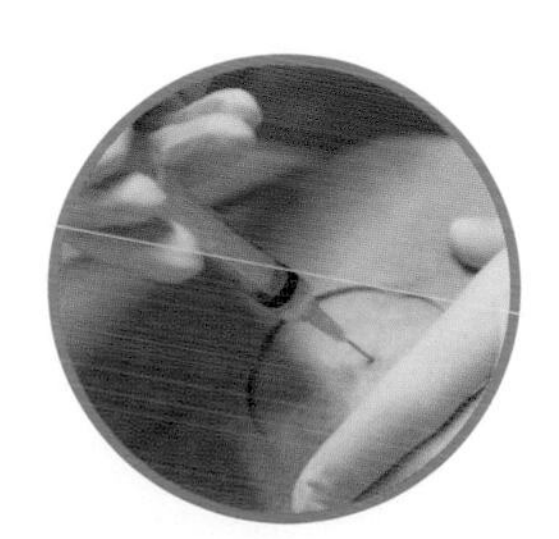

Brimming with creative inspiration, how-to projects, and useful information to enrich your everyday life, Quarto Knows is a favorite destination for those pursuing their interests and passions. Visit our site and dig deeper with our books into your area of interest: Quarto Creates, Quarto Cooks, Quarto Homes, Quarto Lives, Quarto Drives, Quarto Explores, Quarto Gifts, or Quarto Kids.

© 2018 Oxford Publishing Ventures Ltd

This edition published in 2018 by Chartwell Books,
an imprint of The Quarto Group,
142 West 36th Street, 4th Floor,
New York, NY 10018, USA
T (212) 779-4972 **F** (212) 779-6058
www.QuartoKnows.com

All rights reserved. No part of this book may be reproduced in any form without written permission of the copyright owners. All images in this book have been reproduced with the knowledge and prior consent of the artists concerned, and no responsibility is accepted by producer, publisher, or printer for any infringement of copyright or otherwise, arising from the contents of this publication. Every effort has been made to ensure that credits accurately comply with information supplied. We apologize for any inaccuracies that may have occurred and will resolve inaccurate or missing information in a subsequent reprinting of the book.

10 9 8 7 6 5 4 3 2 1

ISBN: 978-0-7858-3500-4

Printed in China

PICTURE CREDITS

The images listed below are in the public domain unless otherwise stated. The original patent diagrams are referenced in the text. Abbreviations: Library of Congress: loc.gov; Smithsonian Institution: si/edu; US National Library of Medicine: nlm.nih.gov.
Internal images: 8 si/edu / 9 brescia.uwo.ca / 10 Getty Images / 11 David Wall/Alamy / 12 George Rinhart/Getty / 13 ignazuri/Alamy / 14 loc.gov / ERproductions Ltd/Getty /15 loc.gov / 17 Avpics/Alamy / 18 Hulton Archive/Getty / 19 Paramount Pictures / 20 nasa.gov / 22 David Taylor Photography/Alamy / 24 shutterstock / North Wind Picture Archives/Alamy / 25 si/edu / nasa.gov /27 Digital Vision/Getty / 28 nlm.nih.gov / 29 magam-safety.com / 30 mit.edu / tulane.edu / 31 shutterstock / 32 melitta-group / shutterstock / 34 loc.gov / 35 cnn.com / 36 si/edu / 37 si/edu / 38 Medicshots/Alamy / 41 uspto.gov / 42 si/edu / molekuul.be/Alamy / 43 Zoonar GmbH/Alamy / 44 twitter.com / 45 Patrick Kovarik/AFP/Getty / 47 joycechenfoods.com / 48 Bettman/Getty / 50 Sergii Moskaliuk/Alamy / 51 Allstar Picture Library/Alamy / 52 Science History Images/Alamy / 54 shutterstock / 55 civilwarsignals.org / 58 Chris Jackson/Getty / 60 Granger Historical Picture Archive/Alamy / Photo 12/Alamy / Lordprice Collection/Alamy / 62 Nataliya Kuznetsova/Alamy / shutterstock / 63 si.edu/ 65 loc.gov / 66 colaimages/Alamy / Chronicle/Alamy / 67 Ugo Bettini 1926 / 68 Bettmann/Getty / 69 Will and Deni Mcintyre/The Life Images Collection/Getty / 70 Nathaniel Noir/Alamy / 71 Enigma Images/Alamy / 72 moodboard/Alamy / 74 William Taylor Family album / shutterstock / 75 sacbee.com / vreseis.com / 76 Jewish Chronicle Archive/Heritage-Images / Science History Images/Alamy / 77 nobeastsofierce Science/Alamy / 78 Phanie/Alamy / 79 newn.cam.ac.uk / 80 Tony Barnard/Getty / 81 Bertrand Rindoff Petroff/Getty / 82 Johannes Jansson / 83 nlm.nih.gov / SciTech Image/Alamy/ 84 dunlopdairy.co.uk / 85 Library of Congress/Corbis/VCG via Getty / 86 Olaf Speier/Alamy / 88 shutterstock / 89 mandyhaberman.com / 91 Johner Images/Alamy / 92 Everett Collection Inc/Alamy / Mattel via Business Wire / 93 Hemis/Alamy / shutterstock / 95 loc.gov / 96 Interfoto/Alamy / 97 Bettmann/Getty / 98 Bettmann/Getty / 100 Bettmann/Getty / 101 Science History Images/Alamy / 102 Chronicle/Alamy / geogphotos/Alamy / 105 Science History Images/Alamy / 106 britishmuseum.org / Coston Stock/Alamy / 107 Susanne Kischnick/Alamy / 108 Simon Belcher/Alamy / Emilio Ereza/Alamy / 109 shutterstock / 111 si.edu / Nina Leen/The Life Picture Collection/Getty / 113 Joshua Yospyn The Washington Post/Getty / Ed Clark/The Life Images Collection/Getty / 115 loc.gov / 117 si/edu / 119 shutterstock / 120 si.edu / 121 Everett Collection Historical/Alamy / 124 imagebroker/Alamy / 126 Everett Collection Inc/Alamy / 131 Science Museum London / shutterstock / 134 Splash News / Alamy / marqueecinemas.com / 137 Lebrecht Music and Arts Photo Library/Alamy / 138 shutterstock / 139 nlm.nih.gov / Science History Images/Alamy / 140 Science History Images/Alamy / 141 Everett Collection Inc/Alamy / 142 Original photo © Davie Gan 500px.com / 144 Denver Post/Getty / 145 NASA Photo/Alamy / 148 ClassicStock / Alamy / 149 shutterstock / 150 Vintage Images/Getty / 151 Granger Historical Picture Archive/Alamy / 153 Sergey Pykhonin/Alamy / 155 Bertrand Guay/AFP/Getty / 156 Granger Historical Picture Archive/Alamy / 157 shutterstock / Nemanja Otic/Alamy / 159 Nic Hamilton Photographic/Alamy / 160 d-lab.mit.edu / 163 loc.gov / popsci.com / 164 FPW/Alamy / 165 Stephen Lovekin/Getty / 166 Keith Dannemiller/Alamy / 167 PJF Military Collection/Alamy / 168 Science History Images/Alamy / loc.gov / 169 shutterstock / 170 shutterstock / 173 Hulton Archive/Getty / 174 Look Die Bildagentur der Fotografen GmbH/Alamy / 175 Interfoto/Alamy / 178 Bob Thomas/Popperfoto/Getty / 180 Bookworm Classics/Alamy / 181 Christoph Furlong/Alamy / 183 Sergei Finko / Alamy / 184 National Cancer Institute / 185 Alain Nogues/Sygma/Sygma via Getty / 186 si/edu / Science History Images/Alamy / 187 Keystone Pictures USA/Alamy / Cover images are all credited as internal pages.